Geoff
& Owen Ryder

DRIVING STEAM LOCOMOTIVES

An Introduction

© Geoff Holland & Owen Ryder 2008

The rights of Geoff Holland & Owen Ryder as authors of this work have been asserted by them in accordance with the Copyright, Design and Patents Act, 1993.

All rights reserved. No part of this publication may be reproduced, stored in a retrieval system, transmitted in any form by means electronic or manual or photocopied or recorded by any other information storage and retrieval system without prior permission in writing from the publishers.

British Library Cataloguing-in Publication-Data: a catalogue record of this book is held by the British Library.

First Printing 2005
Second Edition 2008

ISBN(13) No. 978-0-9549330-2-9

Published in Great Britain by:

The Siân Project Group
Zandvoort, Smithy Banks, Holmrook, Cumbria. CA19 1TP
www.sianprojectgroup.com

Layout and Design by Camden Studios, Rode, Somerset.

Printed and Bound by Cambridge Printing Ltd.

PLEASE NOTE!
In this book the authors and publisher are only passing on knowledge; if you are inspired to take control of a steam locomotive, be aware that your safety, and that of others, as well as the wellbeing of the locomotive, is your responsibility.

Contents

Chapter			Page
Introduction			1
Chapter 1	Railways – A Few Basic Facts		2
Chapter 2	Steam Locomotive Types		3
	2.1	Wheel Arrangements	3
	2.2	Carrying Fuel and Water	3
	2.3	Cylinders, Frames and Wheels	4
Chapter 3	The Career of a Steam Locomotive Driver		5
	3.1	Cleaner	5
	3.2	Fireman	5
	3.3	Driver	5
	3.4	Other Posts	6
	3.5	Preserved Railway Practices	6
Chapter 4	How a Steam Locomotive Works – The Basics		7
	4.1	Parts of a Locomotive	7
	4.2	How it Works	8
Chapter 5	How a Steam Locomotive Works – in more Detail		11
	5.1	The Boiler: – Fire, Water and Pressure	11
	5.2	The Cylinders	14
	5.3	The Valve Gear	16
	5.4	The Importance of the Chimney and Blastpipe	18
	5.5	Lubrication and Bearings	19
	5.6	The Brakes	20
	5.7	The Controls	21
Chapter 6	How to Drive a Steam Locomotive		27
	6.1	Raising Steam and Preparation	27
	6.2	Moving Off Shed and Joining the Train	29
	6.3	Starting a Train	30
	6.4	On the Move	31
	6.5	Stopping a Train	33
	6.6	Locomotive Care During the Day	34
	6.7	Disposal of the Locomotive at the End Of The Day	34
Chapter 7	Maintenance and Overhauls		37
	7.1	Tube Cleaning	37
	7.2	Boiler Washout	37
	7.3	Brake Adjustment	37
	7.4	Firebar Replacement	37
	7.5	Lubrication of Other Parts	38
	7.6	Overhauls	38
Chapter 8	Driving Miniature and Model Locomotives		39
	8.1	Driver and Fireman	39
	8.2	Water Supply	39
	8.3	Raising Steam	39
	8.4	Water Feed	40
	8.5	Everything Happens More Quickly!	40
Appendix	Siân and *The Siân Project* Group		41
Index			43

Inside the cab - just add the crew

An engine man is a driver, mechanic and manager, considered by his passengers as a celebrity, by his boss as a trustworthy employee and looked upon by railway enthusiasts with envy.

Introduction

The origins of this book lie in a set of course notes that were produced in 2002, as briefing material to be sent out to participants in advance of 'Footplate Experience' days that were being run by the publishers.

Like so many briefing notes, the idea behind them was to ensure that the people participating in the course gained the maximum benefit from their time with the course tutors, in this particular case by having some understanding of the mysteries of preparing, firing and driving a locomotive, before they commenced the day's instruction. It also provided an introduction to the vocabulary used by railwaymen.

The most frequent comment received in relation to these notes was that they genuinely succeeded in their intention, and it would be a good idea if they could be made available to a wider audience. We have thus revised the notes to remove the specifics of the particular railway and locomotive involved in our courses, and provided additional material to give an introduction to railways and steam locomotives in general. Since steam locomotives come in many sizes, and with numerous construction detail differences, further material has been added to help the reader's understanding of these matters.

Fig. 1
Driver under instruction on the Ravenglass & Eskdale Railway.

Chapter 1
Railways – A Few Basic Facts

The basic principle of railways is that of a steel wheel running and guided on a steel rail, such that the rolling resistance is minimised, and the vehicles running on the railway do not have to be steered by the driver. The earliest railways were guided by a flange on the side of the rails, and are now referred to as plateways. Motive power was by horses or manpower. By 1800 or thereabouts, railways utilising flanged iron wheels running on iron rails were in existence, and the very first steam locomotives were beginning to be experimented with. The prime means of specifying a railway is by its gauge, which is the dimension between the inside faces of the top part of the rails.

Passenger carrying public railways have been built with gauges from as little as 15", up to as much as 7 feet. Below 15" gauge is the world of miniature and model railways. The main U.K. railway system is built to 4' 8½" gauge - normally referred to as "standard gauge". This rather odd choice of gauge appears to stem from a decision by George Stephenson to build his first railways on a gauge similar to the track of road carts. Standard gauge is used in numerous other countries worldwide, though it is by no means universal. Russian railways use 5' gauge, and several countries in southern Africa collectively have a huge network, all on 3' 6" gauge, otherwise known as 'Cape gauge'.

Most railways also have a defined loading gauge, which is the maximum height and width permissible for vehicles on that railway, such that they will fit tunnels, bridges, station platforms and the like. Typically, U.K. standard gauge has a loading gauge height of 13' and a maximum width of 9'. North American railways, built with the benefit of what had already been learnt in the U.K., are on standard gauge, but with a loading gauge height of 15' and width of 10' 6". This may not sound like much extra, but it represents a 35% increase in the loading gauge cross-sectional area, and results in the use of much larger locomotives and carriages.

A further limitation on the design of railway vehicles is the maximum permissible axle loading of a particular railway. This is dictated by the cross-sectional size of the rails, and the sleeper spacing to bear the load. It can also be limited by the strength of bridges the railway passes over. U.K. standard gauge generally supports about 20 tons per axle, whereas main routes in the USA can support 30 tons per axle.

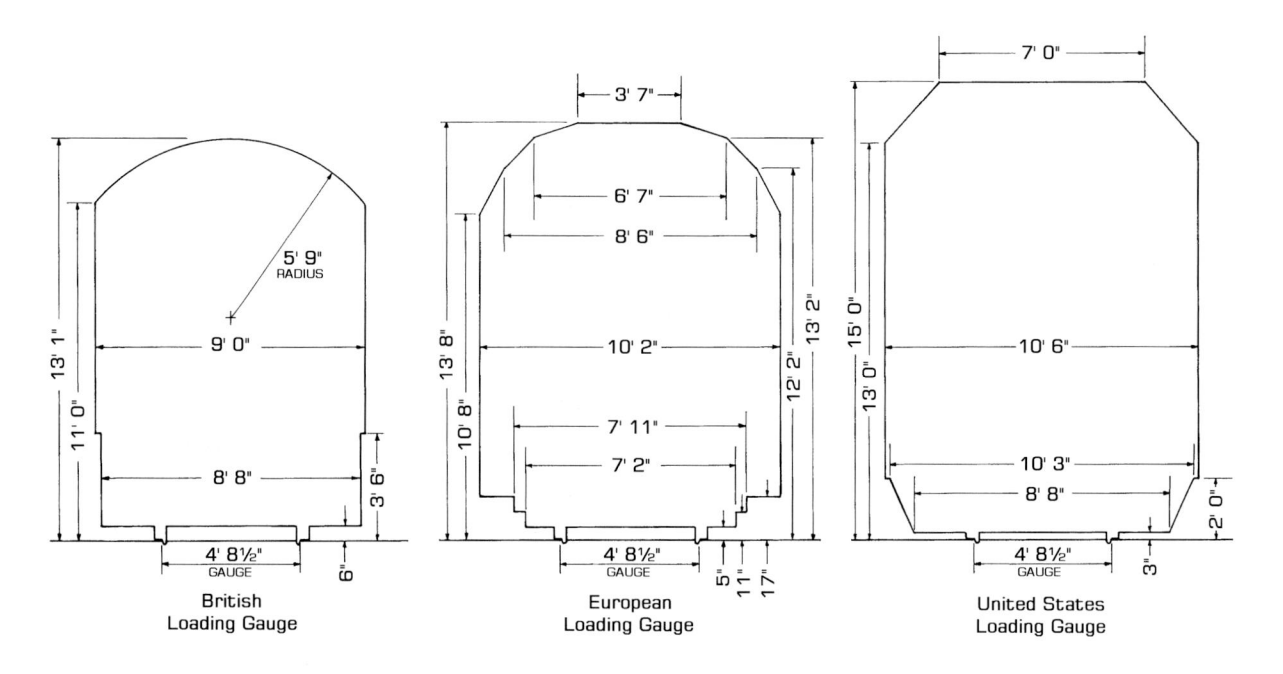

Fig. 2
Typical loading gauges.

Chapter 2
Steam Locomotive Types

2.1 Wheel Arrangement

In general steam locomotives are described firstly by their wheel arrangement. Since a steam locomotive usually has its driving wheels grouped on a rigid chassis, there is a limitation on the wheelbase length in order for it to be able to negotiate curves in the track. Thus, as steam locomotives of greater size and power were needed, it became necessary for some of the weight to be carried on non-driven wheels that could be arranged to have side movement to cope with the curvature of the track, and beneficially provide a means of guiding the driving wheels into the curves.

The notation used to describe wheel arrangements in the U.K. is usually a set of three numbers - firstly the number of non-driven wheels at the front, then the number of driving wheels, and finally the number of non-driven wheels at the back. (The wheels supporting the tender, if one is attached, are not included.) Hence, if a locomotive has four non-driven wheels at the front, six driving wheels, and two non-driven wheels at the back, it is referred to as a 4-6-2.

Some of the wheel arrangements also have names, which can be used in place of the numbers. The origins of the names stem from a variety of sources, but commonly originate from the railway company that first built a particular wheel arrangement. For example a 4-6-2 is commonly referred to as a Pacific, and a 2-6-0 as a Mogul. See Fig. 3 for a fuller listing.

Steam locomotives with more than one group of driving wheels are a specialised area, as they usually require the chassis carrying the driving wheels to be able to articulate to negotiate the curves of the track. They thus need flexible pipework to get the steam from the boilers to the cylinders. This would be relatively easy with modern materials, but 100 years ago, the pipework would have all been in steel and copper with spherical joints to allow for the movement required. The names for articulated locomotives often have their origin in the name of the person or locomotive building company who invented, and often patented, a particular arrangement; examples include the *Garratt*, *Fairlie*, *Mallet* and *Kitson-Meyer* types.

2.2 Carrying Fuel And Water

Steam locomotive fuel is usually coal, but other fuels that can be reasonably burned in the firebox have been used, including oil, wood, coconut husks, and sugar cane residue (bagasse). A small narrow gauge

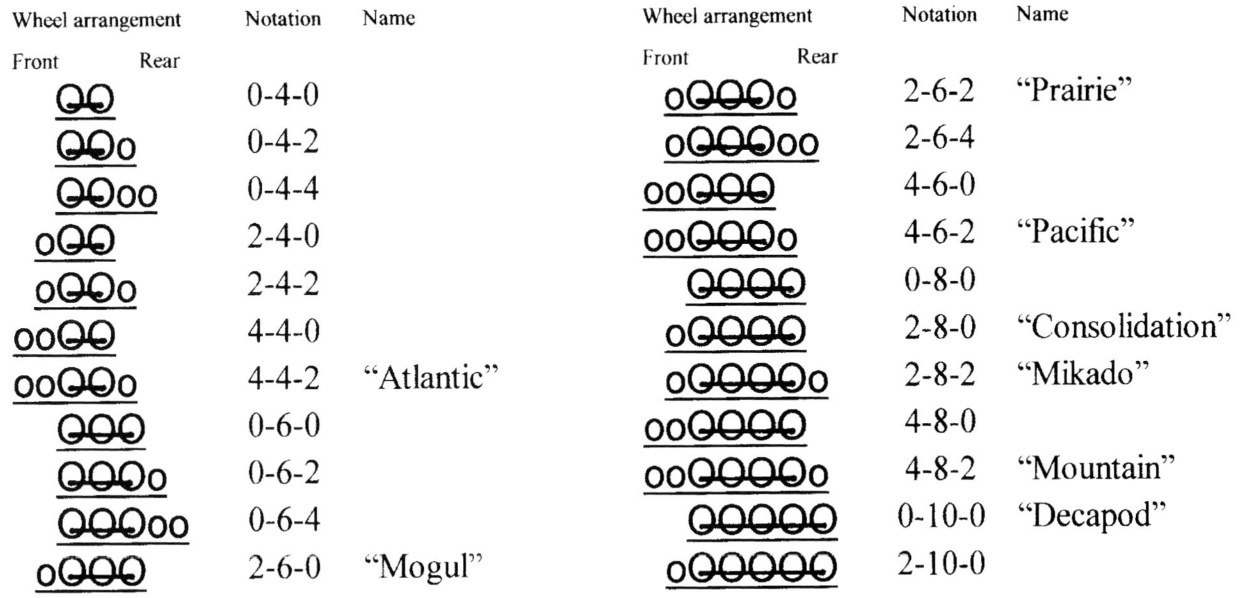

Wheel arrangement	Notation	Name
Front Rear		
○○	0-4-0	
○○o	0-4-2	
○○oo	0-4-4	
o○○	2-4-0	
o○○o	2-4-2	
oo○○	4-4-0	
oo○○o	4-4-2	"Atlantic"
○○○	0-6-0	
○○○o	0-6-2	
○○○oo	0-6-4	
o○○○	2-6-0	"Mogul"

Wheel arrangement	Notation	Name
Front Rear		
o○○○o	2-6-2	"Prairie"
o○○○oo	2-6-4	
oo○○○	4-6-0	
oo○○○o	4-6-2	"Pacific"
○○○○	0-8-0	
o○○○○	2-8-0	"Consolidation"
o○○○○o	2-8-2	"Mikado"
oo○○○○	4-8-0	
oo○○○○o	4-8-2	"Mountain"
○○○○○	0-10-0	"Decapod"
o○○○○○	2-10-0	

Fig. 3
"Whyte" wheel arrangement notation.

locomotive may carry a few hundredweight (200 kg) of coal, and a large British express locomotive, up to 10 tons. A steam locomotive also uses a lot of water; 100 gallons (450 litres) would be sufficient for a small narrow gauge locomotive, and up to 5,000 gallons (22,500 litres) for a large British express locomotive. Larger American locomotives would carry in excess of 24 tons of coal and 22,000 gallons of water.

The next common division in locomotive classification is whether the locomotive unit carries the fuel and water, in which case it is referred to as a *tank locomotive*. If the fuel and water are carried on a separate, but permanently attached vehicle known as a tender, the locomotive is inevitably called a *tender locomotive*. Tank locomotives are further sub-divided according to the style of their water tanks:-
  Side tank – tanks stood on the running boards each side of the boiler,
  Saddle tank – a single curved tank astride the top of the boiler,
  Pannier tank – a pair of tanks slung along the sides of the boiler,
  Well tank – single tank located out of view between the locomotive's frames.

2.3 Cylinders, Frames & Wheels

Further details of a locomotive's construction are often used to describe its type. These include the position and number of driving cylinders. The cylinders may be located between (inside) the frames, or outside the frames or, if the locomotive has more than two cylinders, some combination of inside and outside will be used.

The position of the locomotive's chassis frame in relation to the driving wheels is also subdivided. An outside framed locomotive has very little of the driving wheels visible, as they are hidden away between the frames. An inside framed locomotive has the chassis inboard of the driving wheels, so that the majority of the wheels are visible. All the combinations of cylinder, frame and wheel positions listed above have been tried in locomotive construction, but by the beginning of the 20th century, most mainline locomotives at least, were being designed with inside frames and outside cylinders. This was to give the best access to working parts that required regular attention for lubrication and servicing.

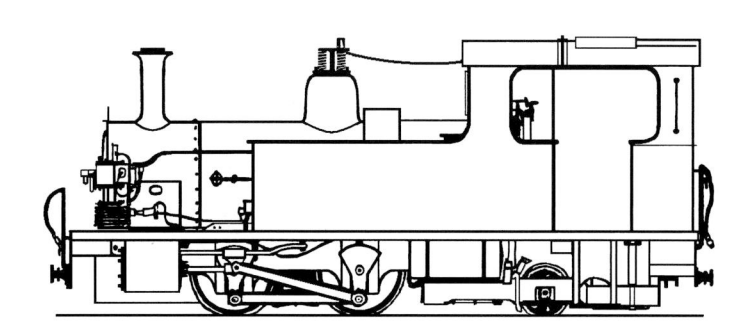

Fig. 4
0-4-2T side tank locomotive with outside frames.

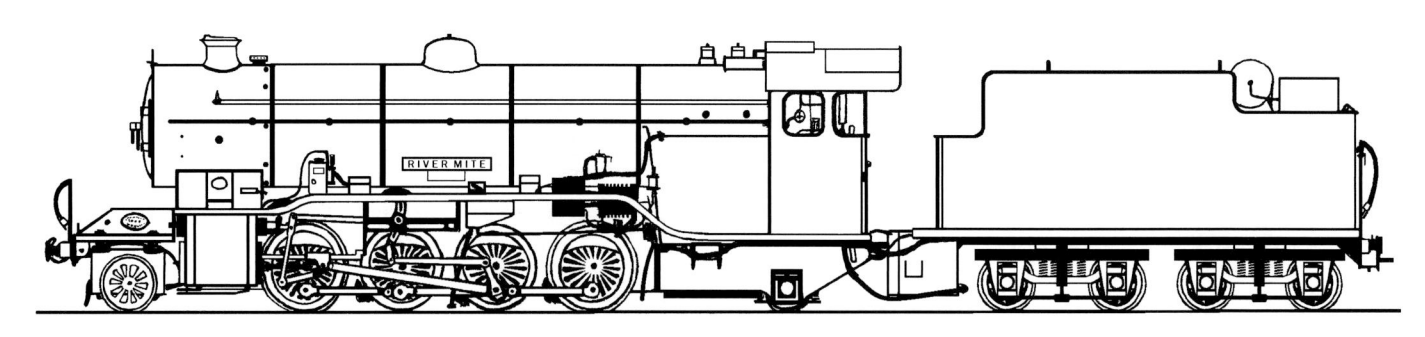

Fig. 5
2-8-2 tender locomotive with inside frames.

Chapter 3
The Career Of A Steam Locomotive Driver

3.1 Cleaner

In the days when steam was the prime mover of the railways, the route to becoming a locomotive driver followed a set path that enabled the aspiring driver to acquire the requisite skills and knowledge along the way.

On leaving school, aged 14 or 15, boys would often join the railways through a family connection. Initially they would be employed as locomotive cleaners and spend their days at the locomotive sheds under the watchful eyes of the shed foreman. Time spent cleaning enabled the boys to learn about the function of the various parts of the locomotives and some further time was spent in learning the railway's rulebook. They would progress to fire lighting, steam raising and helping prepare the locomotives for their work, and the arduous disposal at the end of a working day, when the remains of the fire, and all the ash, had to be cleaned out.

Fig. 6
Fireman on a Southern Railway S15 4-6-0.

3.2 Fireman

From the age of about 17, the cleaners would be sent out with a regular fireman, and taught how to fire "on the road". They would have a basic examination in locomotive operation and railway rules, and would become passed cleaners. This would enable them to be used as firemen when there were no regular firemen available, so they would then build up experience to enable them to be promoted to fireman. Once in grade as a fireman, they would initially fire locomotives on local freight turns, progressing to passenger train firing and eventually becoming a passed fireman.

Becoming a passed fireman involved a similar progression to that of passed cleaner, whereby the experienced fireman was taught to drive and had further examinations in the railway's rules and gave practical demonstrations of competence to the relevant inspector. Quite often a driver and fireman worked as a team for several years and, as the driver was responsible for the fireman, help in progression was usually forthcoming The passed fireman would then be called on to drive local trains when there was a shortage of drivers.

The age at which a fireman could become a driver in the U.K. varied with different railway companies, but could be as early as 25 and as late as 40. It must be remembered that the fireman might have to shovel up to 10 tons of coal in a shift, so it wasn't in a railway company's best interests to allow fit young men to actually drive a train! Far better to have experienced older men for that job.

3.3 Driver

To become a driver, the passed fireman would have already done quite a bit of driving. In addition to being able to drive the locomotive, the driver must also have a detailed knowledge of the routes he will work over. The location and layout of every signal, station, junction, gradient and siding must be learnt. A driver is examined and must sign that he has the knowledge of a route before he can drive it unsupervised. This is known as "signing for the road". He must have had sufficient experience to be able to drive in all weather conditions, day and night, for rain, frost, snow or wet autumn leaves can make a huge difference to how quickly a train can be started and slowed down. There were also further, and more detailed, examinations on the rulebook, knowledge of locomotive operation, and a practical

Fig. 7
*The driver of 0-6-2 tank locomotive "Birch Grove"
preserved on the Bluebell Railway.*

competence demonstration for the examining driving inspector.

On a steam locomotive, the driver's years spent as a fireman would enable him to spot when a fireman wasn't doing things correctly, and to provide help and advice as required. There are times when the steam locomotive driver is given a bit too much credit for the success of a run. In reality, provided that the fireman is able to manage the boiler to provide the right amount of steam when it is needed, a driver has a less arduous task, albeit he is the one ultimately responsible for the safety of his train. Far more driving skill is required when the locomotive does not steam well, or, of course, when driving conditions are difficult.

3.4 Other Posts

There was no room in the footplate progression for individuals who could not make the next grade, so they were weeded out and given less safety critical jobs such as station porters, fitters' mates and boiler tube cleaners in the sheds, or they joined the track maintenance gangs. A railway in steam days was a people intensive industry, and loyal staff could usually be found another post. Some drivers would progress on up the management chain to become shed foremen, or firing and driving inspectors. Medical fitness was critical for footplate staff, including a thorough eye test, for which glasses were often not permitted on the grounds that steam and rain would obscure them. Older drivers who could no longer pass the medical examination were taken off main line work, and put on shunting duties or moving locomotives around the sheds.

3.5 Preserved Railway Practices

On today's preserved railways, progression of the footplate staff largely follows a condensed version of the old style driver's career. Since the longest preserved railway is only about 20 miles, route knowledge is easier to pick up. Most preserved railways have volunteer drivers and firemen, who will have started to work on a railway much later than 15 years old. Age limits for cleaners, firemen and drivers vary from railway to railway, and can be much more flexible, dependent on aptitude, and the amount of time a volunteer can put in. Some railways do set a minimum attendance that must be met to maintain qualification as a driver or fireman. Physical fitness and being able to pass an eyesight test are still essential, though the wearing of glasses or contact lenses is now permitted. However, not only is it no longer necessary to start as a 15 year old boy, being female is no longer a bar to becoming a steam locomotive driver, and some preserved railways have been able to roster an all female train crew on occasion.

Quite often the route to becoming a driver is through helping to restore a locomotive, and it is not unusual for the locomotive crew on a preserved railway to own, or have a share in, the locomotive they are actually operating. If, having read this booklet, you are inspired to take the matter further, contact the preserved railway you are interested in volunteering on - most will make you welcome and explain their operating regime to you. If you would like to try driving and firing a steam locomotive, many preserved railways offer courses, with experienced tutors, lasting from a day to a week, though at a cost compared to starting at the bottom and volunteering.

Be warned - it can be addictive and many a day or weekend's footplate experience has led to regular volunteering, and sometimes even becoming part-owner of a locomotive.

Chapter 4
How A Steam Locomotive Works - The Basics

4.1 Parts of a Locomotive:

The diagram below shows a narrow gauge tender locomotive. The parts labelled are generic to most steam locomotives, though dependent on size and design they can be configured differently. For example, the safety valves (5) could be positioned on top of the firebox or next to the whistle, and the feed water delivery pipes (10) could go into the boiler on top of the boiler barrel, or at the firebox end.

1) Smokebox – This is where the exhaust gases from the fire come out of the boiler.

2) Chimney – This allows the mixture of smoke and steam out of the smokebox.

3) Boiler – This contains hot water and steam under pressure, heated by a fire.

4) Dome – The dome is placed high on the boiler, to collect the steam away from the water level where it will be driest, and least likely to carry water into the cylinders.

5) Safety Valves – These are placed on top of the dome on this locomotive but, as noted above, could be mounted elsewhere on the boiler, providing they are above the steam space. If the boiler pressure reaches the maximum allowed, these valves automatically open and release the excess pressure. They close again when the pressure reduces.

6) Firebox – The firebox is the part of the boiler that contains the fire.

7) Whistle – A noisy device to draw attention! Many railway companies had their own design of whistle, so you could recognise just by the whistle tone which company a locomotive belonged to.

8) Injector – This is used to inject water from the water tank into the boiler. Most locomotives are equipped with two injectors.

9) Blowdown Valve – This is used to allow sediment to be discharged from the bottom of the boiler.

10) Feed Water Delivery Pipe and Non-Return Valve – Where boiler feed water is injected into the boiler. The non-return valve is usually referred to as a *Clack Valve* from the noise it makes.

11) Leading Pony Truck – The leading wheels help guide the locomotive into corners.

12) Cylinders – The steam enters the cylinders, which contain pistons. The pistons push the wheels round via the motion. (14)

13) Cylinder Drains – Opening these releases water in the cylinders which results from condensed steam.

14) Motion & Valve Gear – The motion connects the

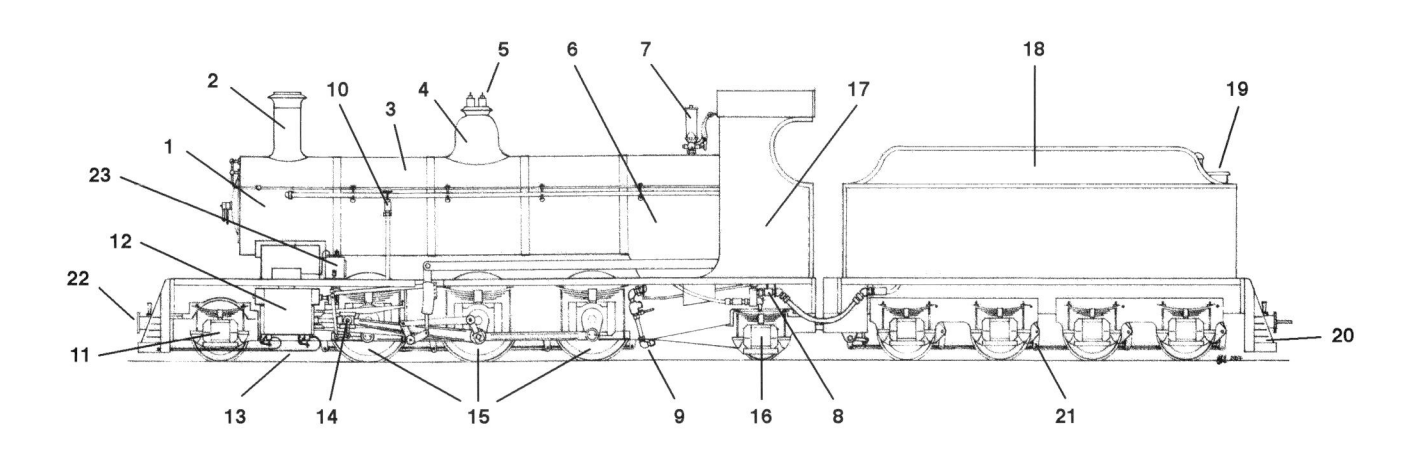

Fig. 8
Diagram showing the main parts of the locomotive labelled.

piston to the cranks and wheels. It also moves and controls the admission of steam to the front and back of the piston in the cylinder.

15) Driving Wheels – These wheels transmit the power to the track.

16) Rear Pony Truck – The rear truck consists of a small pair of wheels on a pivoting frame. It helps support the back of the locomotive and guide the locomotive round corners.

17) Cab – The cab on a locomotive offers the crew some protection from the weather. The floor area of the locomotive cab, on which the crew stand when operating the controls is known as the footplate, hence being "on the footplate" refers to riding in the cab of a locomotive. The term stems from the fact that early locomotives did not have a cab, so originally "on the footplate" was the only accurate description of where the locomotive crew stood.

18) Coal Space – The tender carries coal or other fuel on top of the water tank.

19) Water Tank Filler – Water supplies are carried in the tender body. A steam locomotive does not usually carry enough water for a working day, and the tank should be replenished as the opportunity arises.

20) Cow Catcher – These were employed, mainly on unfenced railways elsewhere than the U.K., to deflect cattle and stop them de-railing the train. In the U.K. and any other country where snow is encountered, locomotives were often individually fitted with snowploughs for the winter months.

21) Brakes – These are present on the locomotive and the tender, and consist of cast iron shoes that rub on the outside of the wheel to slow it down. Steam locomotives can have either air or vacuum powered brakes for stopping the train. If they are fitted with train air or vacuum brake equipment, this can also be linked to the locomotive's own brake system. A handbrake for parking is usually fitted as well.

22) Couplings – There are a variety of couplings used to couple railway vehicles to form a train. These include centre hook and links with side buffers, or a combined centre buffer and coupling, or a centre automatic claw coupler known as a 'buckeye'.

23) Mechanical Lubricator – This is a small tank of oil, which contains a pump driven off the valve gear, to put oil into the cylinders. Lubrication makes the locomotive run smoothly and reduces wear. Without it, the locomotive could seize up.

4.2 How it Works:

4.2.1. The Boiler

The boiler is a large tube of water with smaller fire tubes that pass through it. A fire in the firebox creates the heat. The hot gases from the fire pass through the tubes, warming up the water, and exit through the chimney. The boiler can be coal fired, but as mentioned earlier, other fuels can be used. It is the fireman's job to make sure the fire has enough fuel and is hot enough to create full steam pressure. When the water reaches 212°F (100°C), it boils and steam is created. As the boiler is heated further the temperature and pressure begin to increase. Typically, at a steam pressure of 200 lb/sq inch, the steam/water temperature is 380°F (194°C), so touching an unlagged boiler surface is not advisable. The water level will drop as steam is used, so the fireman uses an injector to inject water from the tender into the boiler. The level must be controlled so that it can be seen in the water level gauges (see Section 4.2.3), indicating that water is covering the top of the firebox.

The driver controls the flow of steam into the cylinders by opening the regulator. This is a valve that lets steam out of the boiler and into the main steam pipe.

4.2.2. The Chassis

The main steam pipe leads to the steam chest, which contains the valves that control which end of the cylinder the steam goes to. If the reverser lever is in forward gear, the steam will go to one end of the cylinder to push the locomotive forward. If the reverser lever is in reverse, the steam will go to the other end of the cylinder to push the locomotive backwards.

The cylinders contain movable pistons. The steam pressure pushes these pistons to the end of the cylinder. The piston is connected to a driving axle by the piston rod, crosshead, connecting rod and cranks. The driven axles are coupled together by coupling rods so that all driving axles are powered.

When the piston reaches the end of its stroke, the valve opens the exhaust port and the exhaust steam is discharged through the smokebox, and up the chimney. The pistons are double acting so they work in both directions. As the steam passes through the smokebox to the chimney, it helps to draw air through the fire.

4.2.3. The Controls

The basic controls are listed below, and the majority identified in Fig. 9. Most are simple in operation, and generic to almost any steam locomotive; however, their position in the cab will vary between different locomotive types.

1) Regulator – Admits steam to the cylinders to make the locomotive move.

2) Reverser – Changes the direction of travel. In Fig. 9, a lever reverser is shown, but some locos have a wheel operated screw reverser.

3) Handbrake – Used for parking and rarely for slowing down. On a tender locomotive, it is usually on the tender (*not shown*).

4) Locomotive Steam Brakes – Give good braking when the locomotive is moving on its own, but not sufficient when hauling a heavy train.

5) Train Brakes – Give good braking when hauling braked trains. Can be either air or vacuum operated.

6) Cylinder Drains – To release condensation from the cylinders.

7) Injector Water Valves (left and right) – For supplying water to the injectors.

8) Injector Steam Valves (left and right) – For powering the injectors.

9) Blower – Used to control draught through the fire when the regulator is closed.

10) Firehole Door – For feeding coal to the fire.

Fig. 9
The cab controls of a narrow gauge locomotive.

11) Boiler Pressure Gauge – Shows steam pressure in the boiler.

12) Water Level Gauge – Shows the water level in the boiler.

13) Steam Chest Pressure Gauge – Shows steam pressure at the cylinder valves.

14) Train Brake Gauge – Shows air pressure in locomotive reservoirs, and in the train pipe on air-braked trains.

15) Whistle – Can be either chain or lever operated.

Fig. 10
STEAM LOCOMOTIVES COME IN MANY SIZES - *the locomotives illustrated here show some of the common gauges; 'Royal Scots' in the 'model' gauge of 3½" in No. 1, and in the 'miniature' gauges of 7¼" & 10¼" in No's 2 & 3. In 15" gauge a German Pacific on the Romney, Hythe & Dymchurch Railway is shown in No 4, 500 mm narrow gauge in No. 5 (Garratt "L.D. Porta on El Tren del Fin del Mundo), 2' narrow gauge in No. 6 (a South African Garratt on the Welsh Highland Railway), standard gauge in No. 7 (a 'Castle' class 4-6-0 heads the crack "Cornish Riviera Express" over the banks of Devon) and broad gauge in No. 8, represented by a preserved 5' gauge Russian P36 class 4-8-4.*

Chapter 5
How a Steam Locomotive Works - in More Detail

5.1 The Boiler: Fire, Water & Pressure

5.1.1. Parts of the Boiler

Boilers come in many different shapes and designs, but most have the same parts. Fig. 11 is a schematic diagram of a boiler with the parts labelled. It is shown with the smokebox still attached as it forms an essential part of the boiler system.

1) Barrel – The barrel is made from steel plate.

2) Steam Space – The part of the boiler above the water, where the steam accumulates.

3) Water Space – The bottom of the boiler contains water. The water level must be kept above the top, or 'crown', of the inner firebox (5a).

4) Water Level Gauge – As it is essential that water is kept above the firebox crown, the level gauges show the crew where the level is. It is important that these function correctly, so normally two gauges are provided for safety.

5) Firebox – The firebox consists of an inner firebox (5a) between the fire and the water and an outer firebox (5b) between the water and the outside. These are kept the correct distance apart by stays (5c) which stop the inner and outer firebox distorting under the pressure. The radiant heat from the fire is transferred via the inner firebox to the water.

6) Fusible Plug – This is a lead (Pb) plug fitted in the firebox crown. Normally the lead is solid and blocks the hole in the inner firebox. If the water drops below the plug, and hence the firebox crown, the lead melts, releasing any remaining water and steam into the firebox where it will (or should) extinguish the fire. Without a fusible plug, uncovering the firebox crown could go undetected. It would soon become too hot, deform and a boiler explosion would occur.

7) Fire Hole – The fire hole is fitted with a door which allows coal to be added, but can be closed so that air is drawn through the fire instead of through the fire hole.

8) Fire Tubes and Superheater Flues – The fire tubes run through the boiler barrel, below water level, between the firebox tubeplate and the front tubeplate. They allow hot fire gases to pass through the water space, transferring more heat, and then into the smokebox. The fire tubes are usually between 1½" and 2¼" outside diameter, and a large boiler can be fitted with up to 200 of them. Superheater flues (8a), if fitted, are located parallel with the fire tubes, and are screwed into the firebox tubeplate.

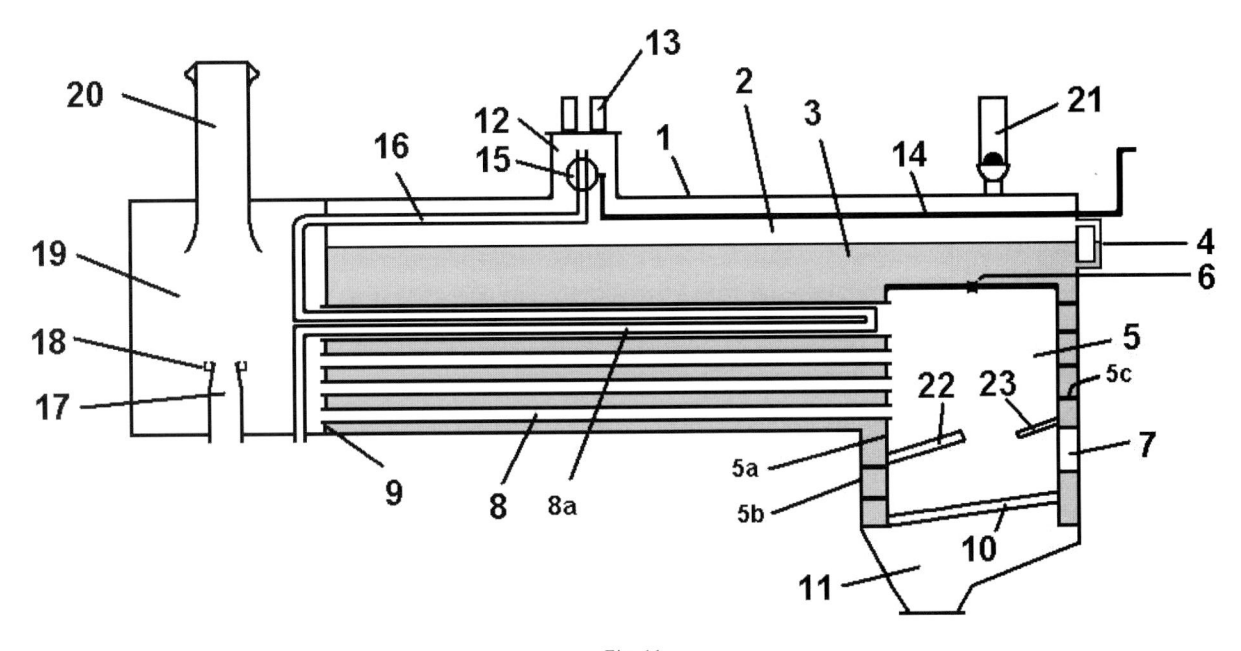

Fig. 11
Schematic drawing of a locomotive type boiler, showing the major parts.

*Fig. 12
A typical locomotive boiler and smokebox, showing layout and construction details.*

They are usually between 3½" and 5½" outside diameter, and contain the superheater elements. These are used to route the steam, on its way from the boiler to the cylinders, into contact with the hot gases, so the steam temperature is raised further than it can be when it is in contact with the water in the boiler. At 250 lb/sq inch boiler pressure, the water and steam temperature is 400°F (205°C). By passing the steam through a typical locomotive type superheater, its temperature can be raised to nearly 750°F (400°C). Superheating the steam allows more work to be extracted from it in the cylinders, albeit at the expense of a more complicated boiler.

9) Front Tubeplate – This plate forms the front end of the pressure vessel.

10) Fire Grate – This comprises a series of cast iron bars to hold the fire in place.

11) Ashpan – This collects the ash falling through the fire bars. It is usually an enclosed box, except for the grate at the top, so that air inlet valves or "dampers" that restrict airflow can be used to control the amount of air to the fire.

12) Dome – The dome is placed high on the boiler to collect the steam furthest away from the water level where it will be driest.

13) Safety Valves – These prevent the pressure rising too high in the boiler. They are set to release steam at the maximum allowed boiler pressure. Usually a boiler will have two safety valves, but up to four are used on larger boilers. More modern locomotives have "pop" type safety valves. These release suddenly, stay open until the pressure has reduced enough, and then they pop shut equally quickly. Such is the sudden nature of the opening, they can startle nearby people or animals. Older style safety valves are more gradual in action, but they tend to remain open much longer and do not bring the pressure down as far. All safety valves are tested to ensure they will release enough steam even when the fire is burning fiercely.

14) Regulator Rod – This rod allows the handle in the cab to control the regulator valve in the dome.

15) Regulator – The regulator valve controls the amount of steam to the cylinders. Numerous designs of regulator valve have been devised, all aimed at providing a smooth, steam tight valve that is easy to operate. The common types are slide valve, poppet valve, or multiple valve. Multiple valve types are often located in the smokebox, where the main steam pipe exits the boiler. Locomotive designers of the past did not have materials like stainless steel and PTFE available to them, which would have revolutionised regulator valve design.

16) Main Steam Pipe – Carries the steam from the dome to the cylinders.

17) Blast Pipe – Exhaust steam from the cylinders passes through the blast pipe and up the chimney. Due to the shape of the chimney and blast pipe, the discharge of steam creates a partial vacuum in the smokebox and this pulls air through the fire. The harder the locomotive works, the more steam goes up the chimney, the harder the draught through the fire and so the hotter the fire gets. If the design of the chimney and blastpipe are correct, and there is the right amount of fire in the firebox, a steam locomotive boiler is able to produce steam at roughly the rate to match the work being done. The final fine balance of steam supply to work rate is achieved thanks to the skill of the fireman, in terms of rate of fuel addition to the fire, setting of the dampers, and use of the injectors.

18) Blower Ring – This is a small pipe with holes that point jets of steam up the chimney. The effect of these is to create a partial vacuum in the smokebox which then pulls air through the fire making it burn more fiercely. The blower is used to create draught when there is no steam going up the blast pipe, and is controlled from a valve in the cab.

19) Smokebox – The smokebox acts as a vacuum chamber. It also collects ash brought through the tubes by the fierce draught.

20) Chimney – This allows the mixture of smoke and

steam out of the smokebox. It is shaped internally to help the exhaust steam from the cylinders create the partial vacuum in the smokebox which draws air through the fire.

21) Whistle – The whistle can be mounted directly to the boiler, or on a manifold, and is controlled by its own integral valve operated by a pull chain or lever.

22) Brick Arch – This is a firebrick construction, spanning the front of the firebox, such that the flame path from the front of the firebox is not straight to the tubes. This results in better combustion, particularly when burning coal, and reduces the extent to which the tube ends are burnt away. It is not necessary on small locomotives, but all standard gauge mainline locomotives will be fitted with one.

23) Deflector - this deflects top-air down to mix with the combustion gases.

5.1.2. Combustion

For combustion to take place there must be fuel (coal in this case), oxygen (from the air) and heat. Coal is put into the firebox through the fire hole door. The air comes from two places; primary air enters through the ashpan under the fire, passes through the fire and allows combustion to take place in the fuel. Secondary air enters through the fire hole door. Primary air is normally controlled by the dampers.

Secondary air is controlled by opening the fire hole door, or a vent in the door on some locomotives.

Coal that has been freshly put on the fire will not burn immediately. Initially, it will give off combustible gases known as volatiles, so secondary air is required for a short period to allow these to burn. Once the gases have been burnt, the secondary air can be reduced.

If the fire is very hot, the ash deposits can melt. When the fire cools, this molten ash sticks together, forming what is called "clinker". If lots of clinker builds up, it blocks the air passages through the fire grate, reducing the draught and thus the fire will not burn as hot as required. It is important that the fireman ensures the fire is clean, and free of clinker, whenever the fire has been burning fiercely.

The limit of firing coal using a man with a shovel is usually reckoned to be when a boiler grate area is about 50 to 60 square feet. The largest steam locomotives built had grates up to 150 square feet, and a mechanised coal firing process is then required, in which a steam powered screw conveyor is used to bring the coal from the bottom of the bunker up to the firehole. Powerful jets of steam are then used to throw the coal to the four corners of the grate. The rate of firing is controlled by the speed of the stoker engine, though it is usually supplemented by the fireman using his shovel to achieve the exact

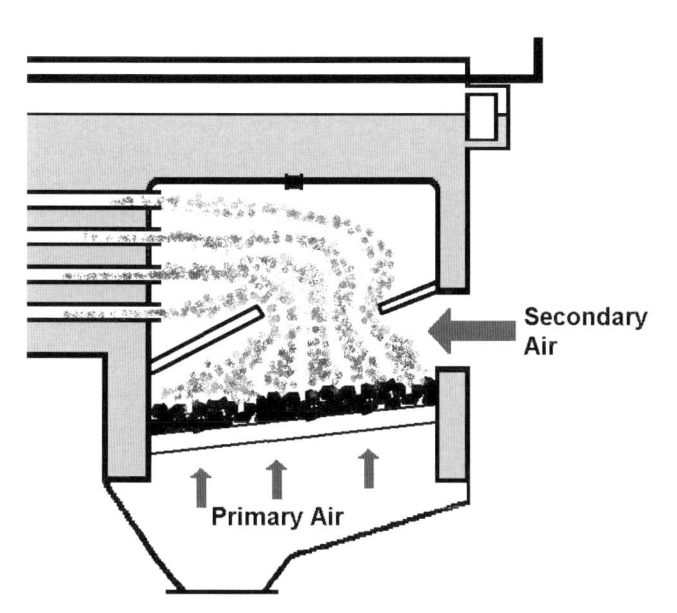

Burning volatile gases released from the coal requires secondary air as well as primary air.

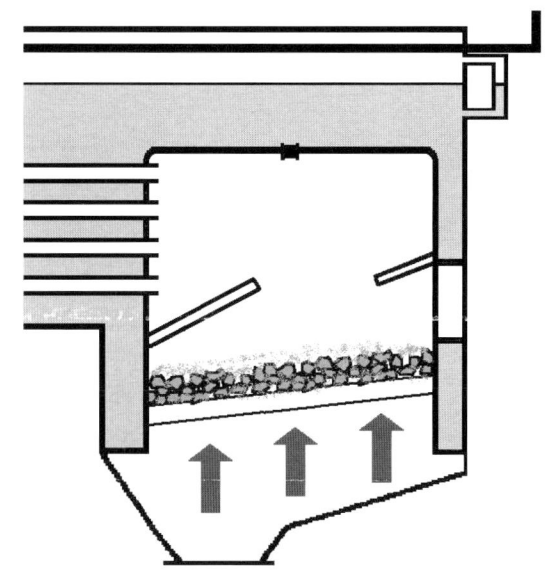

Once the volatile gases are burnt, the coal should be combusting brightly, and primary air drawn through the grate is needed.

Fig. 13
Combustion within the firebox.

fire shape required. Stoker firing is less fuel efficient than hand firing, and involves the expense of its installation, so it is only used when the size of the grate makes it necessary.

Oil firing of steam locomotives has been used where coal is not readily available. The oil is usually burnt in the form of atomised droplets sprayed at pressure through the burners located in the bottom of the firebox. Atomising is achieved using compressed air during lighting up, and using steam once the boiler is up to pressure. Unless modern automation is provided, it is still necessary to have a fireman to control the burners, as they must be carefully adjusted to match the required steaming rate.

5.1.3. Injecting Water

The injectors are used to put more water into the boiler. Even though the injected water has steam condensed in it, its temperature is only about 140-160° F (60-70°C), and hence well below the boiler water temperature. Thus, as it enters the boiler, the water in the boiler is cooled and, since there is a direct steam temperature/pressure relationship, the pressure also reduces. If the fire is burning fiercely it can input enough heat to maintain the boiler temperature and pressure, which is the equilibrium a good fireman is aiming for. When there is less demand for steam, water can be injected to stop pressure from rising and lifting the safety valves. Dependent on design, injectors will stop working when the pressure drops below around 40 psi.

The injectors usually use steam from the boiler to power them, but there are also specialised exhaust steam injectors that can operate using some of the steam that would otherwise be discharged via the blastpipe.

Before the injector was invented, locomotives used to be equipped with water pumps driven off an axle. This was all very well, but they absorbed power and didn't allow water to be put into the boiler when the locomotive was stationary and, unless a feedwater heater was fitted, the water was cold, considerably disrupting the temperature/pressure relationship.

Steam powered feed pumps are occasionally used, normally in conjunction with feedwater heaters.

In order to put water into the boiler, the injector must deliver water at a higher pressure than the boiler pressure. The theory behind their operation is more complicated than can be covered in this book, but a simple explanation is included below.

First the water is turned on (Fig. 15). It will simply exit through the drain, as it is at atmospheric pressure and not able to push into the boiler on its own.

Next, the steam is turned on (Fig. 16). The steam condenses as it hits the cooler water, and the resultant suction effect draws in more water. The high velocity of the condensed steam carries the water across the drain gap and into the delivery cone. As the water goes through the delivery cone, the increasing cross sectional area means it slows down and the kinetic energy (velocity) is converted into increasing pressure. The resulting high pressure is enough to open the non-return valve (called the "clack") and inject the water into the boiler.

5.2 The Cylinders

The main steam pipe goes from the regulator, via the superheater (if fitted) to the steam chest, which is usually located above the cylinder. The steam arrives at the valve, the position of which will dictate

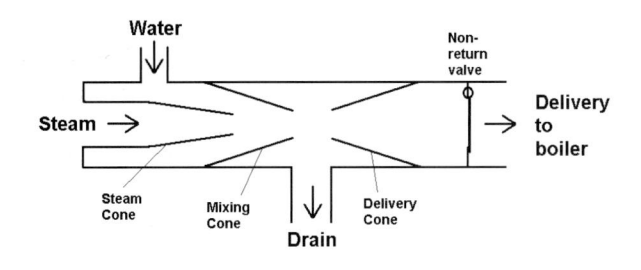

Fig. 14
Cross section of an injector.

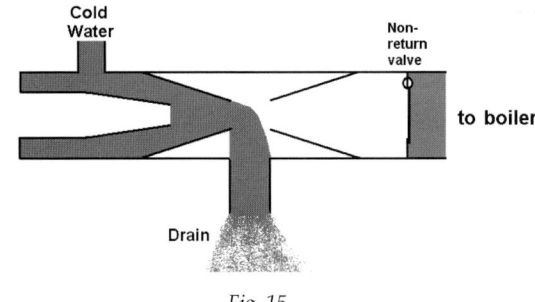

Fig. 15
Injector - water on.

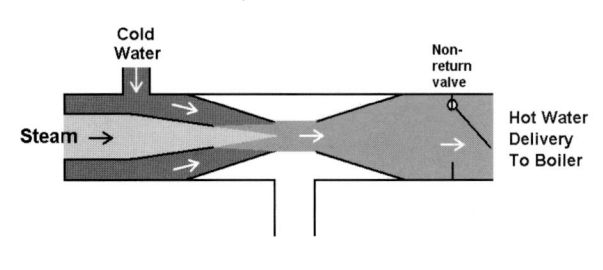

Fig. 16
Injector - water and steam on.

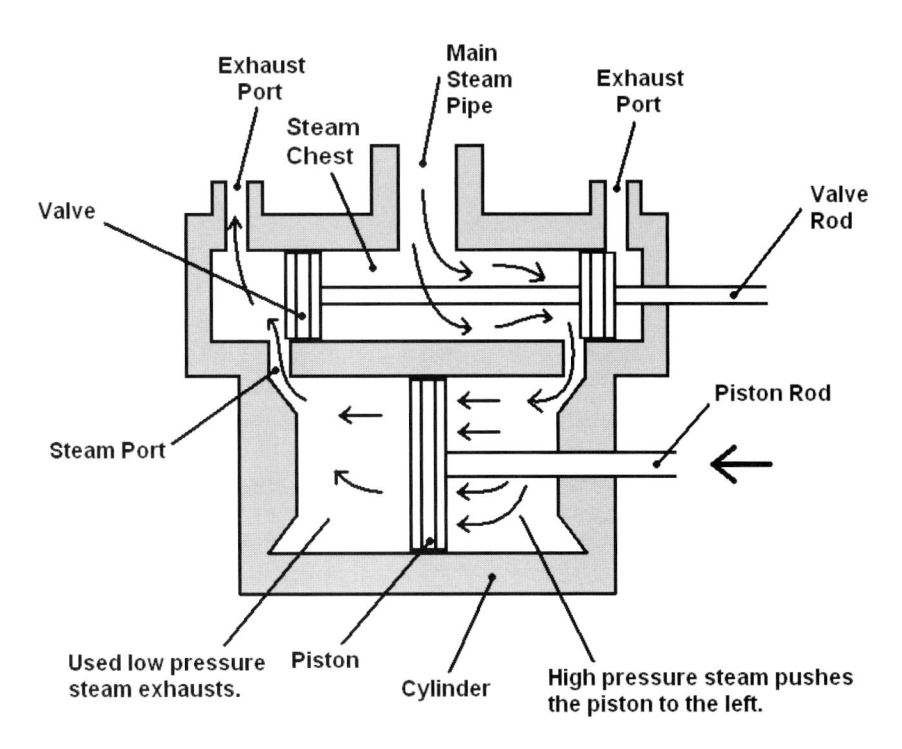

Fig. 17
Cylinder with piston type valves - piston moving to left.

the end of the cylinder the steam goes to next.

In Fig. 17, as the steam enters the cylinder, it pushes the piston to the left. Part way through the stroke, the valve closes the steam port, allowing the steam in the cylinder to expand. As this is happening, the steam on the other side of the piston is allowed to escape through the exhaust port to the blast pipe.

At the end of the stroke, the valve moves to uncover the inlet port to admit steam to the left of the piston and let the used steam from the right of the piston

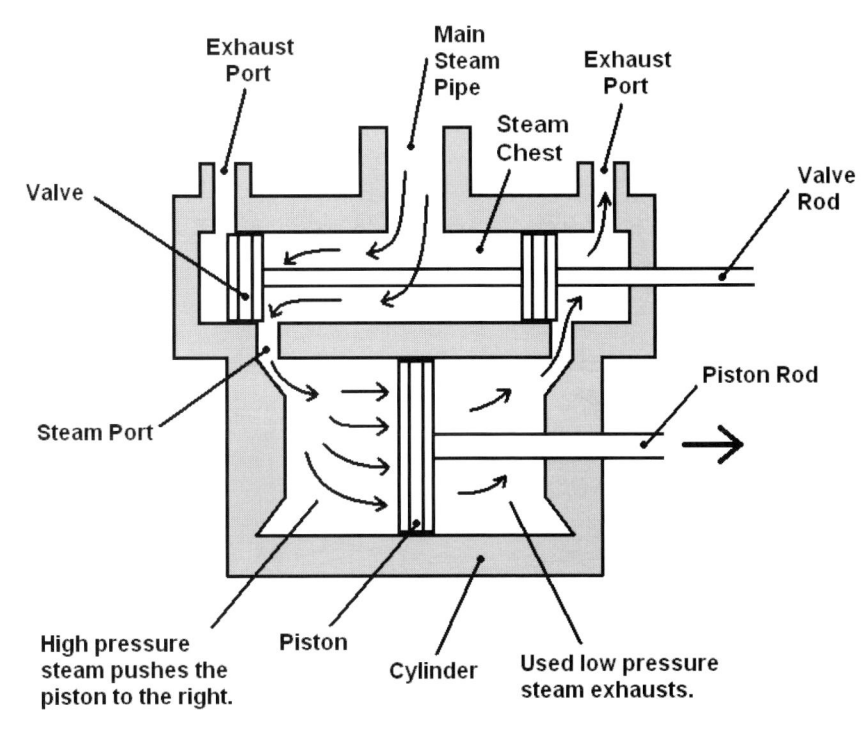

Fig. 18
Cylinder with piston type valves - piston moving to right.

escape to the blast pipe. The piston will then move in the other direction, to the right, as shown in Fig. 18.

The cranks on one side of a two cylinder locomotive are set at 90° to those on the other side so that one of the pistons will always be part way through its stroke and able to move the locomotive. Other angles are used on locomotives with three or four cylinders.

The valves on most recent steam locomotives are piston type with inside admission, where the live steam enters at the middle of the valve, and the exhaust steam goes out via the ends. Outside admission piston valves have the live steam at the ends, and exhaust steam goes up the middle. Slide valves were used before piston valves were developed. These are flat, and slide on a flat face over the two steam ports; the live steam is on top, and the exhaust steam is directed via a cavity in the valve face. Poppet valves more closely resemble the system used in a 4 stroke internal combustion engine, where there are separate inlet and exhaust valves driven by rotary cams.

5.3 The Valve Gear

This book does not intend to describe everything that is needed to design or set valve gear, but it is useful for the driver to have some knowledge of how the valve gear works. The valve gear is a collective name given to the series of levers that control the valve events. There are many types of valve gear, a selection of which are described in this book.

The basic principles of valve gear are to open and close the steam cylinder's valves in the right sequence to drive the pistons, to provide forward and backward movement of the locomotive. The cut-off of the valves (the point at which the valves stop steam entering the cylinder) needs to be variable, so that expansion of the steam in the cylinders can be achieved, and the whole process needs to be under the driver's control.

Expansive working is beneficial because, at the end of the stroke, there is less pressure remaining in the cylinder. This is more efficient than having the full pressure in the cylinder for the full stroke, as the energy would be wasted by exhausting it up the chimney.

5.3.1. Stephenson Valve Gear

Stephenson valve gear is the one most normally used when the valve gear is between the frames of the locomotive. Each valve is provided with two eccentrics, one set at the full forward position and the other at the full reverse position. The ends of the eccentric rods are joined by a curved link, from which the drive to the valve was taken. By raising and lowering the link, the position of the drive from the link to the valve can be varied from full forward, through mid gear, to full reverse. Positions closer to mid gear provide shorter cut-offs of steam at the valves, and hence expansive working. The raising and lowering of the Stephenson link is controlled from the cab via the reversing lever or screw. Fig. 19 shows a typical layout for Stephenson valve gear.

5.3.2. Walschaert Valve Gear

Walschaert valve gear is principally used to drive piston valves on outside cylindered locomotives, though as it can be mounted between the frames, it was also favoured for three and four cylindered locomotive designs. Its drive is taken from a single eccentric, up to a curved swinging link, called an

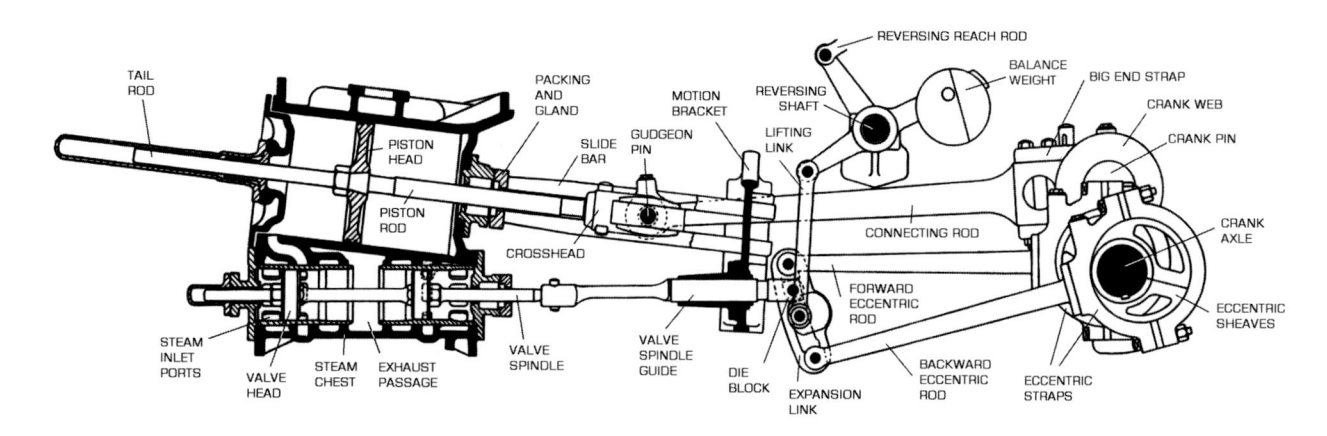

*Fig. 19
Stephenson valve gear.*

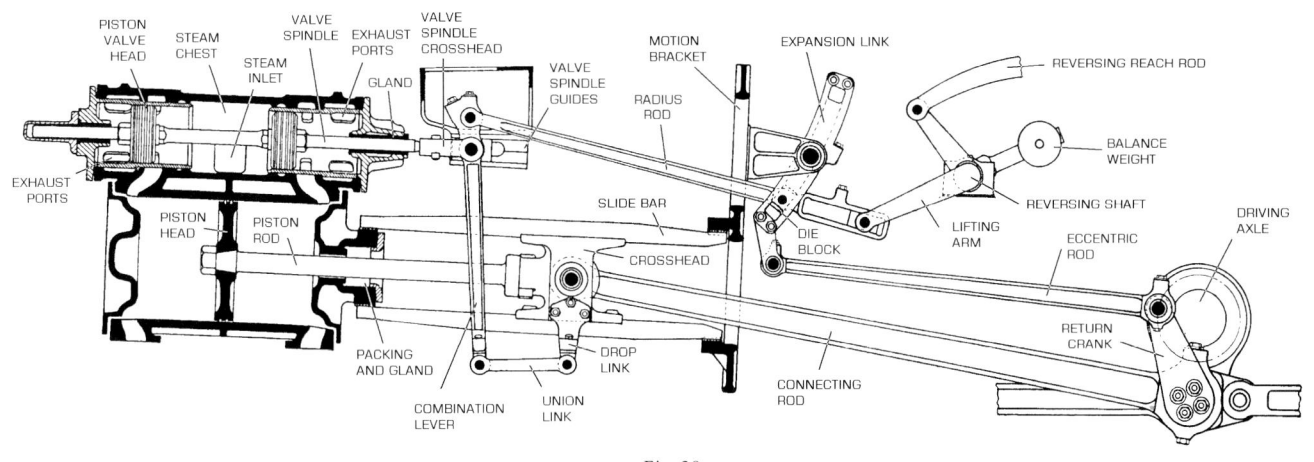

Fig. 20
Walschaert valve gear.

expansion link. The expansion link is mounted on a central pivot, and the drive to the valve is taken from it via the valve rod. If the drive to the valve is taken from below the expansion link pivot the locomotive will move in one direction, and it will move in the opposite direction if the drive is taken from above the pivot. Full forward and full reverse gears are taken from the extremities of the link. As the valve rod drive is taken from nearer the pivot, shorter valve movements are provided to give expansive working. A second drive to the valve is taken from the piston crosshead, and this is combined with the valve rod drive via the combination lever. The combined movements provide quick opening of the valve, followed by slower closing. The position of the valve rod in the expansion link is controlled by the driver using the reverser. Fig. 20 shows a typical layout for Walschaert valve gear.

5.3.3. Joy Valve Gear

Joy valve gear derives the valve motion from the connecting rod. The vertical movement of the connecting rod is taken to an inclined expansion link. The inclination of this expansion link is dictated by the position of the reversing lever and converts the vertical motion to horizontal motion. A steeper inclination of the link gives more valve travel and inclining the link in the opposite direction reverses the direction of valve travel. Fig. 21 shows a typical layout for Joy valve gear.

5.3.4. Twining Valve Gear

This efficient but rare valve gear, designed by Ernest Twining, is included here as it is fitted to the locomotive owned by the publishers, and is visible in a number of the photographs in this book.

Twining valve gear incorporates features of more popular types of valve gear such as those designed by Joy and Walschaert. In it, the valve motion is taken from an eccentric return crank and the direction of travel is set by the inclination of the

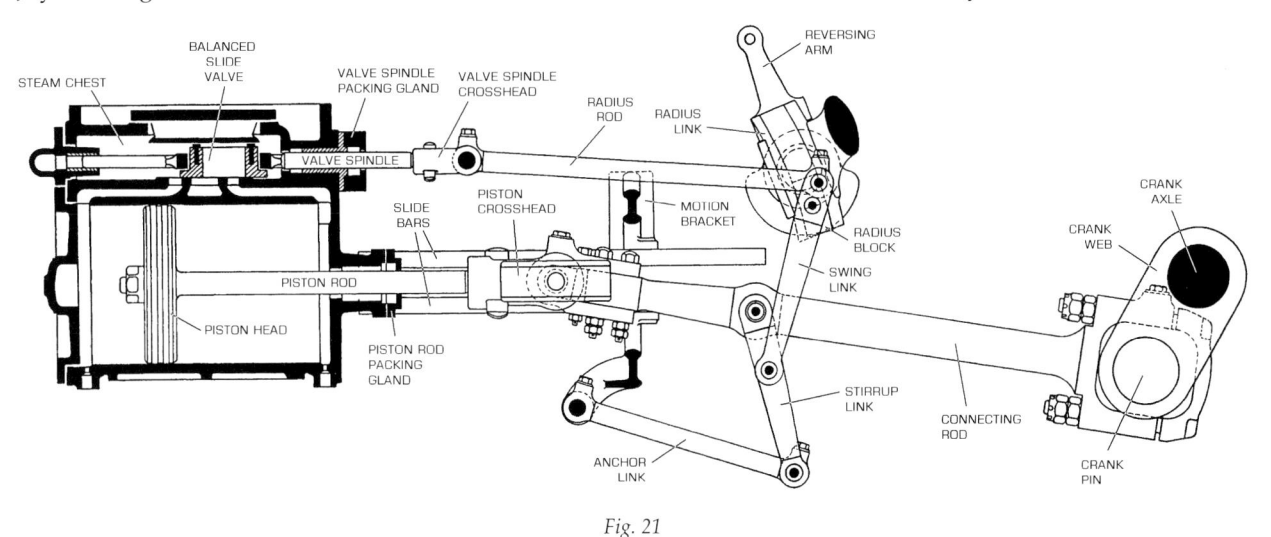

Fig. 21
Joy valve gear.

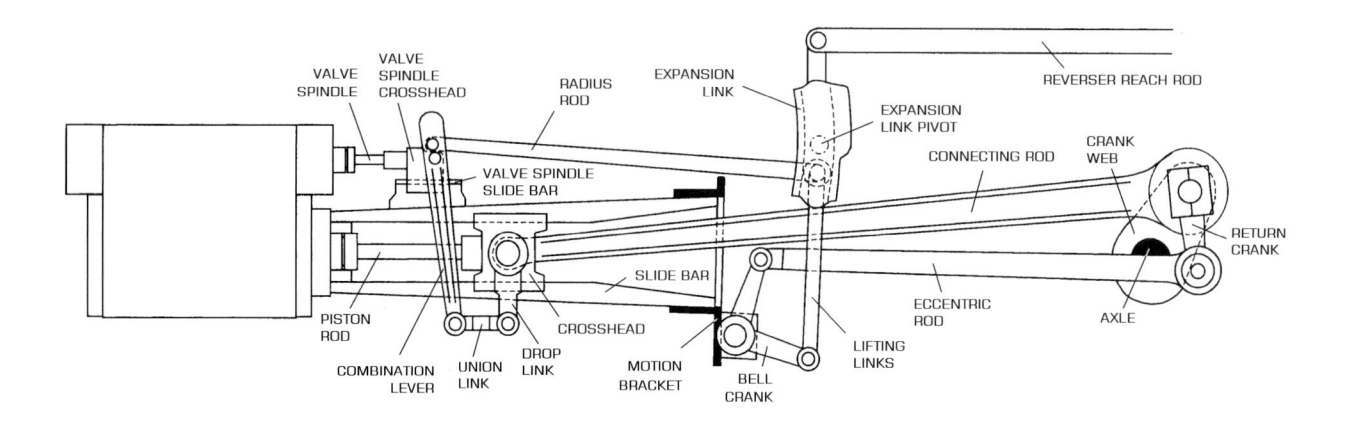

*Fig. 22
Twining valve gear.*

expansion link. This is directly connected to the reversing lever in the cab by means of the reach rod. As the wheels rotate, the lifting links rise and fall in the expansion link. If the expansion link is leaning forwards, the valve events will be such that the locomotive will be powered forwards. If the expansion link is leaning backwards, the locomotive will be powered backwards.

If the expansion link is leaning forwards only slightly, the valve will travel only a short distance, reducing the amount of time that the inlet port is open and thus reducing the cut-off and allowing more efficient working. Fig. 22 shows the layout of Twining valve gear.

5.4 The Importance of the Chimney and Blastpipe

The importance of the chimney and blastpipe in simple form was amply demonstrated by Stephenson's *Rocket*, which is reckoned to have won the Rainhill trials as it was fitted with this arrangement, when its competitors were not.

For many years, the performance of a locomotive was said to depend on its ability to boil water, and it was considered the boiler was the heart of the locomotive. However, some of the most respected steam locomotive engineers, from about 1930 onwards, established that the true heart of a steam locomotive is the blastpipe and chimney.

Whilst most steam locomotives have a single round chimney, there is a limitation to this, in that on larger locomotives the height of the chimney is limited by the loading gauge of the railway. There are ideal ratios for the length and diameter of a chimney and its position in relation to the blastpipe.

In order to get nearer to achieving this, some larger locomotives are fitted with two, and sometimes three, smaller diameter chimneys. These are usually mounted in line in a single casting.

The design of the blastpipe and chimney inside the smokebox can be as simple as a nozzle, lined up with a tube. For best performance, greater complication is required, involving multiple nozzles and conical tubes, which enable the best smokebox vacuum to be obtained with minimum backpressure in the

*Fig. 23
View of smokebox showing main steam pipes, blastpipe, blower ring and front tubeplate. (The horizontal pipe entering from centre left is the exhaust steam pipe from an external air compressor)*

Fig. 24
View of the smokebox of a very modern three cylinder compound locomotive, showing the three Kylchap exhausts used. The main steampipe to the cylinders curves in on the right.

cylinders. The names of the designs originate from their inventors, or combinations of their names. The combined work of Kylälä and Chapelon provided the Kylchap exhaust used on the world speed record locomotive *Mallard*. Other notable exhaust system designers included Giesl, and the Lempor blastpipe designers Messrs. Lemaitre and Porta.

5.4.1. Smoke Deflectors

One result of fitting a multiple blastpipe and chimney is that the exhaust from the chimney does not get thrown clear of the locomotive with the required vigour to ensure it doesn't obstruct the driver's view of the line ahead. This sometimes happens even on single chimney locomotives. Smoke deflectors have to be fitted to create an updraught round the front of the locomotive. These take the form of rectangular plates fitted each side of the smokebox.

5.5 Lubrication and Bearings

The lubrication of early steam locomotives was a rather haphazard affair, as quality mineral oils had not been invented, so animal fats were used. The application of the lubricant could be as crude as stopping the locomotive and dropping a ball of fat down the blastpipe to lubricate the cylinders. Once refined mineral oils were available, matters improved considerably. Axleboxes and rotating parts could be fed from oil reservoirs known as oil boxes, using worsted wool wicks acting as siphons, which gave a steady, controllable flow of oil. Cylinders were fed with oil by use of displacement or hydrostatic lubricators, which consisted of a closed vessel filled with oil, which was fed with steam. As the steam condensed into water, it displaced the oil, which was then fed to the cylinders. Mechanical lubricators were first used on railway locomotives in 1908. Thereafter these superseded hydrostatic lubricators. A mechanical lubricator has a pump, or series of pumps inside an oil reservoir, these being driven by linkage to the locomotive valve gear. Apart from feeding oil to the cylinders, mechanical lubricators can also be used to feed oil to axle bearings, and various parts of the motion and valve gear. Mechanical lubricators are generally mounted on the running boards of a locomotive.

Various grades of oil are used on steam locomotives, the principal ones being bearing oil for rotating and sliding bearings, and cylinder oil for the conditions inside the steam cylinders. Steam engine oils are not the same as motor vehicle engine oils and neither should be used for the opposite purpose. Cylinder oil comes in grades to suit the steam temperature, dependent on whether the steam is used straight from the boiler (saturated), or further heated after leaving the boiler (superheated). If a locomotive uses superheated steam, it is often necessary to mix the cylinder oil with a small quantity of saturated steam before it is put into the cylinders to prevent it from being broken down to carbon. The mixing devices built into the cylinder lubrication system are called atomisers. These spray the oil onto the sliding surfaces of the valves and cylinders.

Fig. 25
Mechanical lubricator mounted on a locomotive's running board, showing linkage to the valve gear.

Fig. 26
Oil cans and grease gun as used on a narrow gauge steam locomotive.

Grease lubrication can also be used on both plain and roller bearings, the grease being in various grades, from soft, for packing roller bearings, to hard for use on valve gear plain bearings.

Bearings on steam locomotives were at the leading edge of bearing design for many years. Initially plain bearings in brass or bronze with steel axles were used, to be later improved with white metal bearing facings. White metal is an antifriction alloy of tin, antimony, lead and copper. Axleboxes were also made from steel with pressed-in bronze inserts. Ball and roller bearings arrived late on the scene for steam locomotives, and were only used extensively after about 1950. Some of the last batches of steam locomotives built had roller bearings on all axles, coupling rods, valve gear joints, and big ends of connecting rods.

5.6 The Brakes

5.6.1. Locomotive Brakes

The brake shoes on the locomotive are usually applied by steam. This uses steam from the boiler to exert force on a piston which, through pull rods, rubs brake shoes against the locomotive wheel treads. A steam brake only acts on the wheels of the locomotive and not on any vehicles being hauled. As steam pressure is required to work these brakes, a handbrake is also fitted for parking purposes.

In the early stages of railway development, trains were only braked by the brakes acting on the locomotive wheels, with possible assistance from a braked vehicle at the rear of the train, under the control of the guard. For slow moving freight trains in the UK, this state of affairs lasted right up to the end of regular steam traction.

Following several disastrous accidents with inadequately braked passenger trains, legislation was introduced in 1889 to force railway companies to provide all passenger carriages with continuous fail-safe brakes. By fail-safe, this means that if a train is parted inadvertently, then all the brakes are automatically applied to bring both parts to rest. By continuous, this means that all vehicles in the train are fitted, and the brakes are connected so they can be applied in unison. The brake system must also be capable of being controlled by the driver, such that it can be used for normal and emergency stopping of the train.

Two systems of providing train brakes were devised - using either vacuum or compressed air. Vacuum brakes are simpler to fit to a steam locomotive, but air brakes are generally acknowledged as better, so the main UK railway system has largely converted to air brakes since the end of steam traction.

5.6.2. Train Vacuum Brakes

Most UK steam railways use vacuum brakes. In this system, all the brake cylinders on the carriages or trucks are connected via a pipe to the locomotive. Flexible pipes with quick release connectors are used between the vehicles. A vacuum is applied via the pipe to both sides of a piston in a brake cylinder, when the brakes are in the released state. If air is admitted to the brake pipe, it can only reach one side of the brake pistons, as a non-return valve is fitted to prevent it reaching the other side. The atmospheric pressure then forces the piston along the cylinder. By linking the piston to the vehicle brake system, the brakes can be applied - usually by brake shoes directly rubbing on the wheel rims. A vacuum brake equipped locomotive has a vacuum ejector, which uses steam blown through special nozzles to suck the required vacuum. The steam used in the vacuum ejector is discharged up the chimney. This is quite

Fig. 27
Most locomotives have one brake block per wheel, but many German locomotives are fitted with two, as on this 23 class 2-6-2.

How a Steam Locomotive Works - In More Detail | 21

recognisable by a hollow roar that the locomotive may make whilst standing still. When a locomotive has just been coupled to a train, it has to create sufficient vacuum to release the brakes on the whole train, so the intensity of the hollow roaring sound will increase as the driver uses a larger vacuum ejector nozzle for a short period. Most steam railways used a vacuum of 10 pounds per sq. inch below atmospheric pressure, equivalent to 21 inches of mercury - inches of mercury (in. Hg) being the usual unit for vacuum on a steam locomotive.

5.6.3. Train Air Brakes

Air brakes, like vacuum brakes, require a pipe all the way along the train to enable control and fail-safe action to be achieved. Each carriage is equipped with a triple valve, a non-return valve and small air reservoir. The train air pipe is filled with compressed air (usually at between 50 and 90 lb per sq. inch), and this is directed via the triple valves and non-return valves into the reservoirs. If the pressure in the train pipe is reduced, the triple valve senses this and is able to direct air from the reservoirs to the brake pistons linked to the carriage brakes. The non-return valves are there to prevent the air in the reservoirs being lost back into the train pipe. An air brake equipped locomotive has a steam powered air compressor to generate the compressed air and reservoirs to store sufficient air to quickly release the train brakes. Sometimes the compressor is mounted out of view, but an air-braked locomotive can usually be identified by the fact that it 'chuffs', even when standing still.

5.7 The Controls

Mount the footplate of any steam locomotive, of any gauge or scale, and nationality and you will find a number of the controls are always immediately identifiable - these are the Regulator, the Reverser, the Water Level Gauges and the Boiler Pressure Gauge. The other controls listed below will also be present in most instances, although not always so obvious. Fig. 28 illustrates the major controls of a *Great Western Railway* express passenger locomotive circa. 1910, Fig. 29 illustrates those of an *LMS* express passenger locomotive of the 1930s, whilst Fig. 30 shows those of a *Union Pacific Railroad* stoker fired 4-8-2 of the same period.

Steam locomotive controls are either of lever type, such as the regulator, reverser, brakes, cylinder drains and dampers, or are control valves which are operated by turning anti-clockwise to open, and clockwise to shut - just like a household tap. There are exceptions to this - water level gauges usually have quarter turn valves, and these are covered in more detail in section 5.7.9. Injector steam valves are also frequently quarter turn valves, or a similar rapid opening valve, although they actually look different to the valves on water gauges.

The major controls which will be found in the cab of most steam locomotives (identified by number on Figs. 28 & 29) are:

1) Regulator
2) Reverser
3) Handbrake (on tender – Fig. 28)
4) Driver's brake handle
5) Cylinder drain control lever
6) Injector water valves (left and right)
7) Injector steam valves (left and right)
8) Blower
9) Firehole door
10) Boiler Pressure Gauge
11) Water Level Gauges (only one – Fig. 28)
12) Train brake pipe vacuum gauge
13) Sanding equipment
14) Steam heating valve
15) Whistle handle
16) Ashpan damper controls

5.7.1. The Regulator

The regulator is kept in the closed position until the driver is ready to move the locomotive. It is usually

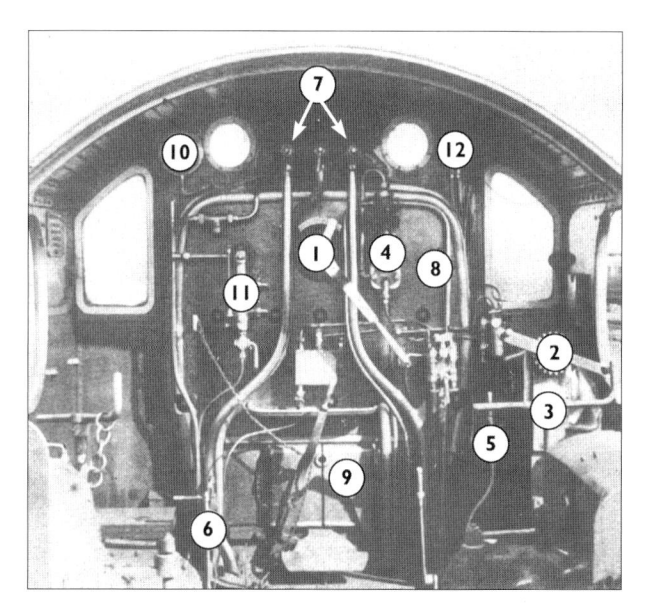

Fig. 28
Cab layout of a Great Western Railway express passenger locomotive of the Edwardian era.

22 | Driving Steam Locomotives – An Introduction

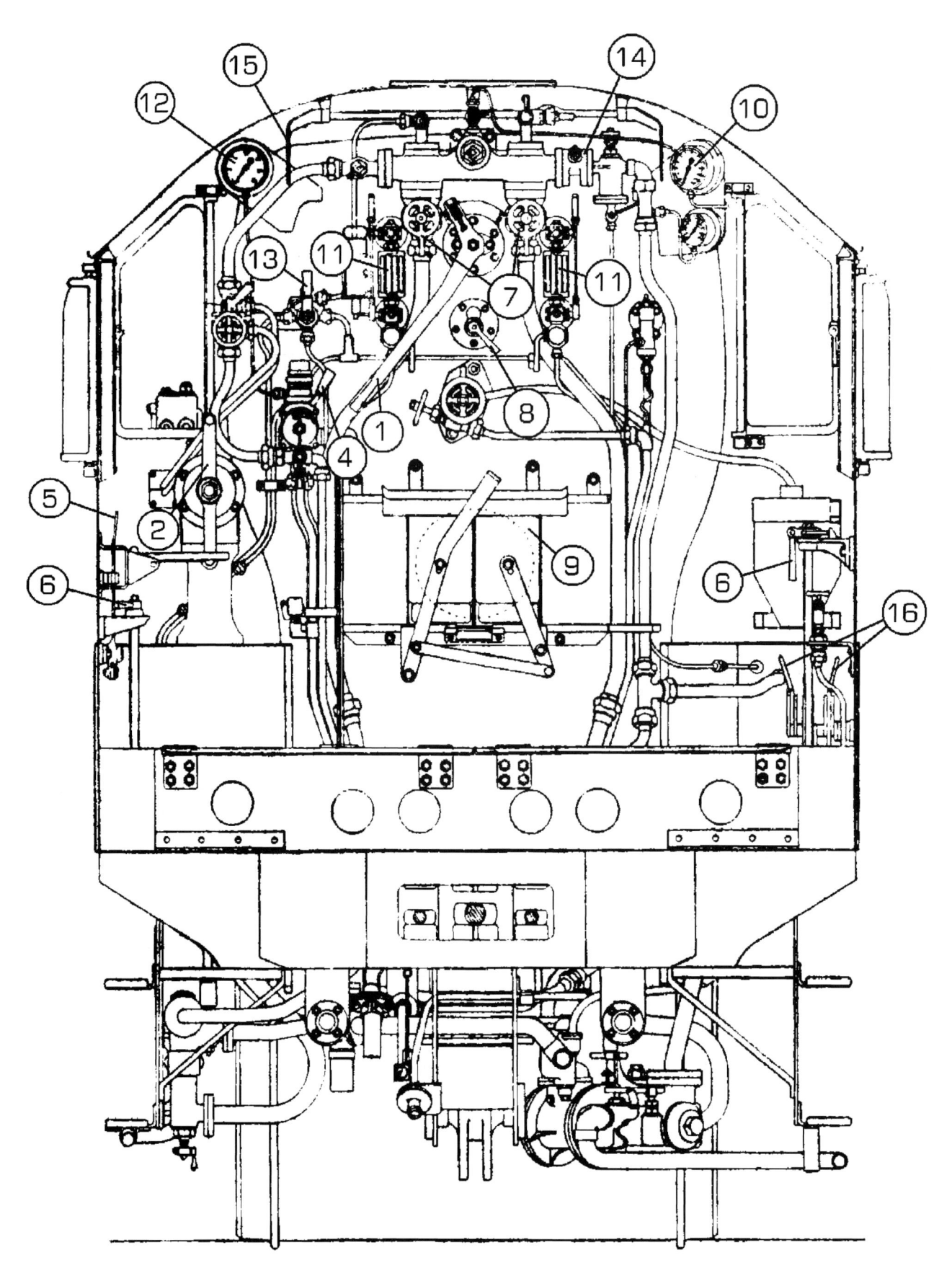

Fig. 29
Cab controls of an LMS standard gauge express passenger locomotive of the 1930s.

a lever operating in an arc across the backhead, or pulled back towards the driver. Generally the pressure of steam on the valve makes the regulator stiff to move initially, but once steam starts to flow, the pressure begins to equalise. After this the regulator is very sensitive and should be opened with extreme caution.

5.7.2. The Reverser

The reverser dictates which direction the locomotive will travel. On smaller locomotives, a lever reverser is used. Due to the forces exerted by the moving valve gear, on larger locomotives a screw reverser is used, utilising a screw and nut arrangement to move the position of the valve gear. A screw reverser is, of necessity, slower in operation, as many turns of the screw are needed to get from full forward to full reverse. Outside the UK, power reversers were commonly fitted. These used steam pistons to quickly push the position of the valve gear and some form of hydraulic circuit to lock it in the selected position.

When setting off, the reverser must be placed in full gear in order to admit steam no matter what the position of the pistons. As the locomotive gathers speed, the reverser can be moved nearer to the central position. This will reduce the valve travel, closing the valve port part way down the stroke (e.g. 25%), admitting less steam and allowing the locomotive to run more efficiently by harnessing the expansion of the admitted steam.

5.7.3. Handbrake

A locomotive handbrake is very much like that on a car, only bigger, and usually applied by a rotating screw and nut, rather than a lever. It should only be used for parking or slow speed braking. It is used to apply force to the same brake system as the locomotive's powered brakes, though on a tender locomotive it only acts on the tender wheels, so it is not very effective for actually stopping.

5.7.4. Locomotive Steam Brake

The steam brake control is by means of a small lever in the cab. The steam brake provides braking force to the locomotive wheels and should be used when slowing or stopping a locomotive. The braking effect will be proportional to the boiler pressure applied to the brake cylinder. If the boiler pressure is low then the braking force will be less. On some locomotives, just one brake handle is present which operates the locomotive steam brake and the train brakes at the same time.

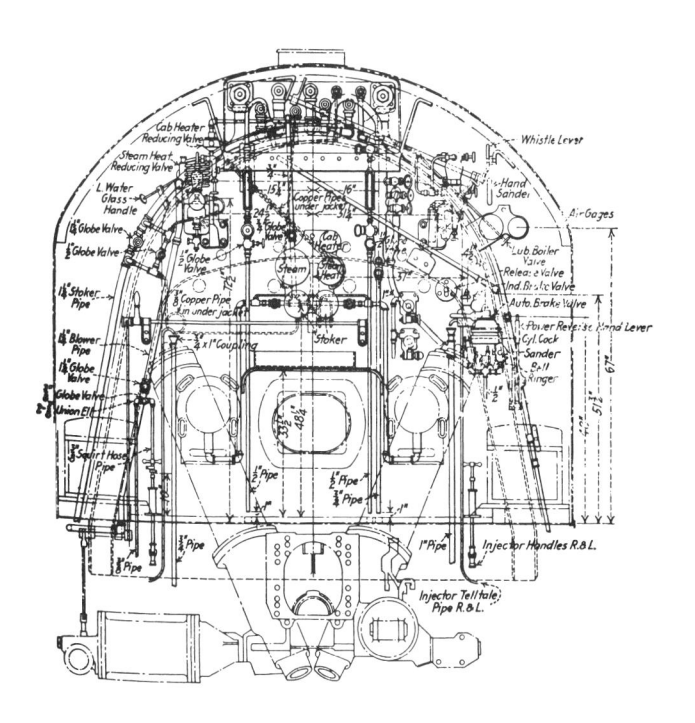

Fig. 30
Cab layout of a stoker-fired American 4-8-2 locomotive

5.7.5. Train Vacuum Brake

If a locomotive is equipped with vacuum brakes, the vacuum must be generated before the locomotive is moved. This is done by turning on the steam supply to the vacuum ejector, and putting the brake control lever to the release position. A vacuum of 21 inches of mercury should be created. To apply the brakes, air is allowed to enter the train pipe by pulling the brake lever. The braking force is proportional to the reduction in vacuum. Hence, if the vacuum is reduced to 15" Hg the brakes will be applied gently, whereas if the vacuum is 5" Hg, braking will be much more effective. After every application of the vacuum brake, whether it is partial or complete, always fully release the brakes to allow the brake chambers to be fully restored to 21" Hg of vacuum.

5.7.6. Train Air Brake

On an air-braked locomotive, the air reservoirs must be at working pressure before the locomotive is moved. The steam supply to the air pump must be turned on and air pressure allowed to build up. The reservoir pressure varies depending on locomotive design but typically can be 100psi and is indicated by a separate pressure gauge.

The brake is operated by the air brake valve, and the train pipe pressure is indicated by a gauge in the cab. Once the brakes are released, air pressure keeps the brakes off. By putting the control lever to the apply position, air can be heard being released from the train pipe. The train pipe pressure will be reduced and the brakes will come on. After every application of the air brake, whether it is partial or complete, always fully release the brakes to allow the train vehicle reservoirs to re-charge.

5.7.7. Cylinder Drain Cocks

To open the drains, there is usually a lever linkage from the cab. Pulling the lever backwards will open the drains. To close the drains, push the lever forwards.

Some locomotives are fitted with steam operated cylinder drains, where steam at boiler pressure is applied to a small piston in each drain cock to hold it closed. A valve in the cab is used to apply and release the steam to this type of drain cock.

5.7.8. Injectors

Two injectors are usually provided for putting water into the boiler. On some locomotives they are referred to as the small and large injectors because they are of different sizes and flow rates. To operate an injector, follow these instructions:-

Step 1) Open the water tap fully so that water exits from injector drain – Fig. 31.

Step 2) Fully open the steam valve. Water will run from the drain more vigorously – Fig. 32.

Step 3) Slowly move the water valve towards the closed position until the injector is heard to "pick up". Check that the water has stopped exiting from the drain – Fig 33. If the water valve is closed too quickly, steam will be blown out of the drain.

To turn the injector off, close the steam valve first, then the water tap.

With experience of the operation of a particular injector, it is possible to set the water valve to the exact setting that will allow the injector to pick up as soon as the steam is fully on.

5.7.9. Water Level Gauges

It is essential to know that the water level shown in the water gauges is correct. Figs. 34 to 39 show how to check that the passages into the boiler are clear from obstruction. Fig. 34 shows the water level gauges in the normal running condition, with all levers pointing vertically. Here, the two taps into the

Fig. 31
Injector water on – slow flow of water.

Fig. 32
Injector steam on – strong water overflow.

Fig. 33
Injector running – no water overflowing.

boiler are in the open position allowing the water level inside the boiler to be shown in the glass. In the vertical position, the drain is closed.

Check that the water bobs up and down to find its level. If the level moves slowly, there may be a blockage. If they do not work properly, declare the locomotive "failed" and rectify the problem. It is safe to run with only one working water gauge glass, but if both are found to be defective, the fire should be dropped immediately, and the boiler allowed to cool before undertaking any repair work.

5.7.10. Blower

The blower increases the draught through the tubes and makes the fire burn more fiercely. It is used to create more steam whilst stationary and should be put on before the fire hole door is opened to stop the flames blowing back into the cab. When turning the blower on, do so slowly. The pipe will be full of condensation, so the blower will initially blow a jet of water from the chimney. When the locomotive is coasting, the blower should be put on prior to entering a tunnel or going under an over-bridge.

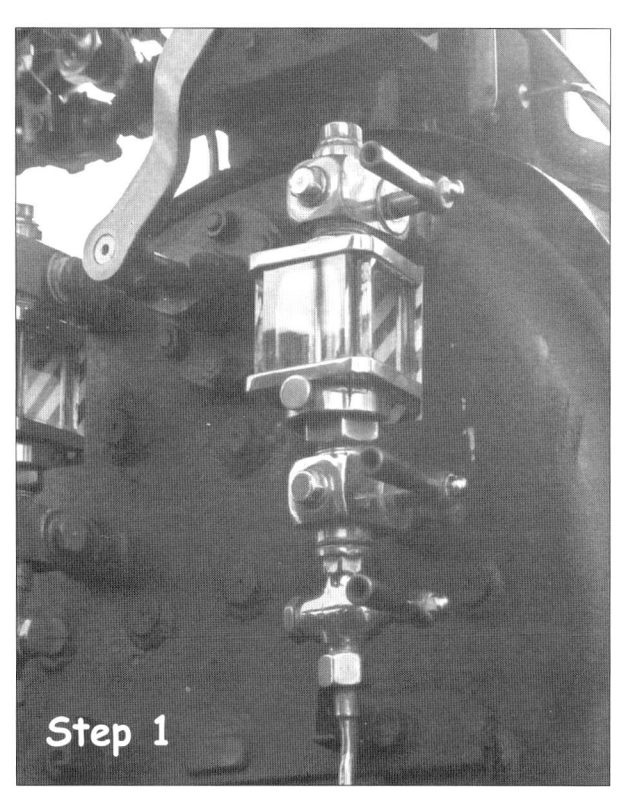

Fig. 35
Close the two taps into the boiler by pulling them horizontal and then open the drain by pulling that handle horizontal as well; no water or steam should come from the drain.

Fig. 34
With all three handles vertical, the taps into the boiler are **OPEN** *and the drain is* **CLOSED**. *These are the normal running positions.*

Fig. 36
Briefly open the top tap and check that steam is blown through the drain. Ensure top tap is closed before proceeding to the next step.

Fig. 37
Briefly open the bottom tap into the boiler and check that water is blown through into the drain.

Fig. 39
Open the drain briefly. When you shut the drain, the water level should bob up and down.

5.7.11. Sanding Equipment

On slippery rail, sand is used to improve the grip between driving wheels and the railhead. It is stored in a sand box and piped to just in front of the wheels. Sanders are either gravity fed, or powered by a steam jet, and controlled by a lever or valve in the cab.

5.7.12. Steam Heating Valve

Trains hauled by steam locomotives use steam from the boiler to heat the passenger compartments. Steam is piped along the train, connected by flexible hoses, and the supply is controlled by a valve in the cab. When nearing the end of a journey, it is very important to turn off the steam heating valve, so that there is no pressure in the system when the locomotive is uncoupled from the train.

Fig. 38
Close the drain (vertical) then open the taps into the boiler (vertical).

Chapter 6
How to Drive a Steam Locomotive

6.1 Raising Steam and Preparation

On arriving for work on a railway, there is usually a requirement to sign on, and to read any special notices that may apply to the job being done.

If a steam locomotive has not been used the day before, steam raising will take some considerable time, as the boiler has to be warmed up slowly to avoid uneven heating and distortion. Even a small, narrow gauge locomotive can take three hours, and a large, standard gauge locomotive can take around eighteen hours to be brought up to working pressure. Some of this time can be spent cleaning the locomotive, which is more pleasant when it is slightly warm.

6.1.1. Positioning

Before the ash is removed or the fire is lit, ideally the locomotive needs to be positioned so that the exhaust is carried away. If the locomotive shed does not have suitable smoke extraction facilities, it is usual to light up outdoors. On smaller locomotives a supply of compressed air may be used to increase the draught, allied to an extension chimney to carry the smoke away.

6.1.2. Emptying of Ash

Open the firebox door and knock any ash or clinker still on the grate into the ashpan. Be aware that the boiler will still be hot if the locomotive was used the day before. Check that the ash is dry. Wet ash is a sign of a leak, perhaps from the boiler tubes or a firebox lap joint. Dry ash will confirm there are no major leaks, but visually check that the tube ends and the fusible plug are dry. Similarly, check that the smokebox is empty and dry.

6.1.3. Check the Water Level

Before the fire is lit, it is imperative to find out how much water is in the boiler. If the firebox crown is not adequately covered by the water, severe damage will occur due to overheating and distortion of the firebox. Open the water level gauge drain and observe that water issues from the drainpipe. Ideally the water level should be about quarter of the gauge glass height.

If the water level is below the gauge glass range, it must be filled until the level is visible. This is usually done by removing a washout plug and filling with a hosepipe.

If the level is too high, it must be lowered, and the excess is drained through the blow down valve. The cold water in the boiler expands more than the boiler itself does as they warm up, so the water level rises. If the water level is too high once the locomotive is initially in steam, there is no room to add more using the injectors, and it is all too easy to have the safety valves lifting before the locomotive has left the shed.

6.1.4. Lighting the Fire

Everyone has their own favourite way of lighting fires. The method described here is that used by the authors. First, lay enough wood to cover the fire grate. Soak some old rags in paraffin, place on the shovel and light. Carefully lower into the firebox, tipping into the middle of the wood. Immediately add some more wood over the burning rags and close the fire hole doors.

Keep the doors closed whenever possible to keep the correct draught flow and stop smoke entering the cab. Check occasionally to see that the fire has enough fuel. When warming a boiler from cold, a small wood fire should be used to heat the boiler through, slowly and steadily. Rushing steam raising will unnecessarily stress the boiler and cause problems later.

Only when a stable wood fire is burning should a little coal be added. Once this coal has started to burn, more can be added. Once the boiler pressure is a little over half working pressure, further coal should not be required until near departure time. A roaring fire is not needed if the locomotive is only going to stay still. Lifting the safety valves whilst still on shed is not only deafening, but represents waste and alerts others to the incompetence of the fireman.

6.1.5. Cleaning

Whilst raising steam, the time can be used to clean the locomotive. Start with brass and copper that will get hot, such as the water level gauges, washout plugs, injector pipework, steam valves, whistle and

Fig. 40
Filling the cylinder lubricators with steam oil.

safety valves. Be sure to shut off the water level gauge cocks before removing the protectors. "Brasso" or similar proprietary polish is used these days. In the past, mixtures of brick dust and oil were used. The bright steelwork can be polished with a metal polish and re-oiled afterwards to stop it going rusty. Heavily tarnished metalwork can initially be cleaned by steel wool or emery cloth in conjunction with some "Brasso".

The problem with keeping a steam locomotive clean is that it is a self-dirtying machine. After a day's service, it will be covered with a mixture of baked on oil, water droplets, soot and ash, plus brake dust, bird droppings and squashed flies. To clean this lot off a nicely finished paint job, without damage, can be quite a challenge. For cleaning the paintwork, all sorts of different techniques are used, each claimed by its user to be the best. These include rubbing with lightly oiled rags, mixtures of paraffin and oil, or water and detergent.

On smaller locomotives a really shiny look can be obtained with least effort by using "Mr Sheen" furniture polish. If you are helping to clean a locomotive for the first time, ask what should be used, because the paintwork becomes conditioned to the cleaning technique, and can be ruined if another technique is used.

The really hot surfaces, such as the smokebox and chimney can be oiled with steam oil to give them a shine. Cleaning of the boiler paintwork is often best left until the locomotive has been moved for the first time, as dirty water from the chimney will ruin initial cleaning efforts. Use clean cotton rags for polishing up brasses and paintwork, then "demote" them to dirtier jobs, eventually disposing of them in a paraffin bin to be used for fire lighting.

6.1.6. Lubrication

Lubricate the moving parts before moving off shed. Grease will be used for any roller bearings and a grease gun will be used on the grease nipples provided around the locomotive. Grease according to the locomotive's lubrication schedule. Wipe off any excess grease otherwise dirt will stick to it and may subsequently get into the bearing system.

Steam oil is a thick, high temperature oil, which is dark green in appearance. It is used in places that need lubricating where steam is present. The cylinder lubricators (mechanical or hydrostatic) will need to be filled. The air compressor (if fitted) will have a lubricator for its steam cylinder, as will the steam brake. Make sure the hydrostatic lubricator, air pump and steam brake shut off valves are shut before undoing the top of the lubricator otherwise it will be under pressure.

Bearing oil is used on the other moving parts, mainly on the valve gear. It is yellow, and thin compared to steam oil. Fill up all the oil pots, checking that the wick is in place as you do so. Also oil any pins that do not have oil pots, and that would benefit from a squirt of oil.

6.1.7. Supplies

Before the locomotive leaves the depot, the driver must check that everything needed during the day is

Fig. 41
Lubricating the motion with bearing oil.

on board. The tender should be full of water and the boiler water treatment liquid should have been added. There must be enough coal in the bunker or tender for the day. Oil cans should be filled before departing. There should be tools for any adjustments that may be needed during the day. If running after dark, the lamps should be in working order and filled with paraffin. If the rail conditions are poor, the locomotive's sand boxes should be checked to ensure they contain an adequate supply of dry sand.

6.2 Moving Off Shed and Joining the Train

There are several checks that should be done before moving the locomotive. These should also be re-checked when taking charge of a locomotive already in steam.

6.2.1. Open All Shut-off Valves

Make sure the injector feeds into the boiler are open, and the steam brake is turned on. Do this by opening the valve (anti-clockwise) as far as possible, and then turning quarter of a turn clockwise again to stop it from jamming open. Check that the steam to the vacuum ejector or air pump is turned on, and the brake reservoirs have the required pressure.

6.2.2. Check the Water Level Gauges Work

As mentioned previously, it is essential to know that the water level shown is correct. See section 5.7.9 for instructions on how to check that water level gauges are showing the correct level.

It is safe to run with only one working water gauge glass, but if both are found to be defective, remove the fire immediately.

6.2.3. Check the Injectors Work

Even if the water level is adequate, check that both injectors operate correctly before leaving shed. See section 5.7.8 on how to operate an injector. If they do not work properly, declare the locomotive "failed" and rectify the problem. A common mistake is to leave the shut-off valve on the boiler shut.

6.2.4. Check the Brakes Work

Apply the steam brake, and check that the brake linkage can be heard to clank as the brakes are applied and released. Set the locomotive moving and check the brake stop it satisfactorily. If they don't work, declare the locomotive "failed" and find out why.

Apply the air or vacuum brake, and check in the same manner as the steam brake.

6.2.5. Clear Cylinders of Condensate

The steam pipes and cylinders will have water in them. This is steam that has condensed from previous use. Water is incompressible and could cause damage if left in the cylinders when the locomotive moves. Warn any bystanders in line with the cylinder drain cocks at the front of the locomotive before clearing the cylinders of this condensate.

With the handbrake fully applied and the cylinder drain cocks open, put the reverser in full forward gear. Open the regulator slightly, but not sufficient to move the locomotive. Observe jets of water emerging from cylinder drain pipework until it clears and only steam passes through. Place the reverser in full backward gear and repeat. Allow the locomotive to roll forward quarter of a wheel revolution and repeat for forward and reverse gear. There should now be no water left in the cylinders.

6.2.6. Moving Off

Blow the whistle to warn bystanders that the locomotive is about to move. Place the locomotive in full gear for the direction of travel, open the regulator so that the locomotive begins to move, and then release the handbrake. Releasing the handbrake before there is steam in the cylinders could cause the locomotive to roll downhill, which may not be the direction intended.

Fig. 42
Condensate blowing from cylinder drains.

6.2.7. Leaving the Locomotive

Before leaving the footplate unattended it is essential that the:-

- **WATER LEVEL IN BOILER IS SATISFACTORY**
- **REGULATOR IS SHUT.**
- **REVERSER IS IN MID-GEAR.**
- **CYLINDER DRAINS ARE OPEN.**
- **HANDBRAKE IS APPLIED.**

Failure to do this could mean the locomotive moves off on its own. It is also a good idea to set the controls like this every time the locomotive comes to a standstill, unless it is to be restarted immediately.

6.2.8. Turntables

The cylinder drain cocks should be open when the locomotive is being placed on a turntable to avoid pressure building up in the cylinders when it is not needed. Check that the turntable is set correctly and locked in position before moving onto it. Run the locomotive onto the turntable slowly, stopping it so that it is in the middle, and preferably balanced. A balanced turntable is easier to turn. Before alighting from the locomotive check that the regulator is closed, the reverser is in mid-gear, the cylinder drains are open, and the handbrake is fully applied.

When turning a locomotive by hand, it should always be pushed round. If pulled, the turntable could run over the operator if they fell. The turntable must be locked in position before moving the locomotive. Only release the handbrake when the locomotive is certain to move in the direction wanted (i.e. when steam is in the cylinders).

6.2.9. Coupling Up

Before approaching the train, make sure the locomotive coupling is compatible with that on the end vehicle. Approach cautiously so as to allow the person making the coupling time to react. The rules of some railways require that the driver stops just short of the train, then moves to couple up. If the locomotive is to be coupled with chimney facing the train, close the drain cocks so as not to scald the person making the coupling. Apply the locomotive brake. Make sure the train brake hoses are connected and move the air or vacuum brake lever to release the brakes. The speed of brake release is dependant on the length of the train. If the brake gauge indicates almost instant release, the chances are a brake hose or valve to the train is not set correctly.

The guard should check that the air pressure or vacuum has reached the back of the train.

Depending on the requirements of the individual railway, train heating and lighting connections may also have to be made.

6.2.10. Permission to Proceed

The driver is not allowed to take the train into a track section until he has permission from the signalman or controller to do so. Dependent on the railway, this may be by the signals, and/or a train token. This token is a brass pole or disc stamped with the section name, and fitted with a hoop to aid passing from person to person. The driver must be in possession of this token before leaving the station. Some railways now use radio telephones, in place of a token system, to issue train control orders. If radios are used, all instructions must be written down by both the driver and the controller.

6.3 Starting a Train

6.3.1. The "Right Away"

The guard is in charge of despatching the train. He will blow his whistle to attract everyone's attention. The whistle is not a signal for the train to go, merely to attract attention. At this stage, the train brakes should be partially released, so a prompt start can be achieved, and the reverser should be in full gear for the direction of travel. The guard will give the "Right Away" by showing a green flag, or raising one arm in the air. Only then should the train be started.

6.3.2. Audible Warning

Blow the whistle to warn bystanders that the train is about to move. If people are standing close to the locomotive, open the whistle valve slowly so as not to startle them.

6.3.3. Starting the Train

Ensure the reverser is in full gear for the direction of travel. Open the regulator slowly until around 25psi is showing on the steam chest pressure gauge. Fully release the brakes and the locomotive should take up the slack in the couplings. The regulator should be opened further, but keep the steam chest pressure gauge below about half of boiler pressure to avoid wheel slip. Slipping reduces the pull on the train, and is potentially very damaging to the locomotive.

If the steam chest pressure goes a bit too high, close the regulator slightly. Time needs to be taken when starting a train. It should not be rushed.

When moving, check that the train is following. Condensate may emit from the chimney as well as the cylinder drains. When this has cleared, the drain cocks can be closed. Once the train is moving, the reverser can be moved back towards mid gear to give shorter valve cut off for more efficient running. The steam chest pressure can be increased if necessary, and the cut-off reduced until running speed is attained. As noted previously, the available grip on the rails can vary considerably, and this must be taken into account if wheel slip is to be avoided.

6.4 On the Move

Once on the move, the driver should be keeping an eye on the following:-

6.4.1. Regulator and Reverser

As the train gathers speed, the cut-off should be reduced until the optimum for the train weight and gradient is attained. Only direct experience with a particular locomotive will enable this to be judged, but effectively the locomotive begins to feel strangled, and ceases to maintain the train speed when too short a cut off is selected. Under the right conditions, the regulator can be fully open and the steam chest pressure will be close to the boiler pressure. At a short cut off, this allows steam to enter only at the very beginning of the piston's stroke, extracting expansion energy from the steam, rather than relying on boiler pressure to push the train along. If a hill is encountered, it will be necessary to move the reverser into longer cut-off - nearer full gear - in order to increase power output for the climb.

6.4.2. Slipping

If the locomotive slips, the exhaust beats will suddenly quicken. Shut the regulator as soon as possible, and reopen carefully so that the driving wheels regain their grip. This needs to be done quickly, otherwise the train can 'bunch up' behind the locomotive, and will need to be stretched out again. Applying a fine stream of sand to the rail just in front of the driving wheels significantly improves the grip of wheels on rail. It does, however, increase the drag of the train, particularly if too much is added. The sand is usually applied using steam sanding equipment, where jets of steam draw the sand from the sandboxes, and fire it at the wheel/rail interface. If sanders are available, but not being used at the time of a slip, don't turn them on until this has been controlled, otherwise the sudden jolt of gaining grip can damage the locomotive.

6.4.3. Priming

Priming is a phenomenon encountered when water is drawn into the cylinders as the locomotive is working hard. This can be caused by low surface tension due to impurities in the water, foaming due to impurities, or because the water level is too high. Signs that priming is starting to occur include a whitening of the locomotive's exhaust at the point when it exits the chimney, a muffling of the exhaust beat and a deluge of water droplets laden with soot. These phenomena indicate excess water is being carried through the cylinders into the exhaust. The regulator valve can become locked by the water rushing through it and the regulator handle may become very difficult to move. Effectively the locomotive can get out of control. When priming starts to occur, close the regulator and open the cylinder drains. If the regulator is locked by the water, put the locomotive in mid gear, then close the

Fig. 43
Safety valves lifting in a station, and coal recently added to the fire - all signs of an inexperienced fireman, but the injector is on, and the train will soon leave. It is 1963, and the steam era is ending, but the steam locomotive continues to give yeoman service.

regulator. If carried into the cylinders, water can severely damage the cylinder castings and/or the connecting rods, as water is incompressible.

Once the locomotive is under control, it will be necessary to proceed at a cautious pace until the water level is brought back to where it should be. If the water in the boiler is contaminated, it may need to be changed by use of the blowdown valve, or by having the boiler washed out.

6.4.4. Keep a Look Out

It is important to keep a look out for dangers on the track, particularly on the first train of the day. Fallen trees, landslides, or acts of vandalism can derail a train. If there are railway personnel on the line, they must be warned by a whistle as the train approaches. They should raise one arm in acknowledgement. The whistle should also be sounded where whistle boards are shown. These are found at level crossings, tunnels and sometimes at stations. Glance back occasionally to check all is well with the train. Carriages swaying violently from side to side means the train is going too fast.

Route knowledge is important; knowing the layout of junctions and which signals apply to the route actually being taken.

6.4.5. Managing the Boiler

The fireman will have to build up the fire prior to departure. The fire will take a period of time to reach maximum heat, so it is not unusual for the boiler pressure to drop initially when setting off. The pressure will recover as the locomotive exhaust draws the fire, and the fire temperature increases.

Once on the run, the fireman needs to maintain the fire at the required level. More coal must be added, evenly distributing it over the fire to match the work rate. If a major climb is coming up, the fire needs to be built up before the climbing starts. Hence it is also important for the fireman to know the route.

Conversely, if the regulator needs to be shut (for a long downhill section or a station stop), the firing should be reduced in anticipation of the stop. Shutting the regulator will stop the use of steam, but if the fire is still burning furiously, the boiler will still be making steam at the same rate and the boiler pressure will increase rapidly. Turning on an injector before shutting the regulator will help keep this pressure rise in check.

On the run, use the injectors to keep the water level at around half a glass, or to stop the safety valves lifting. In an ideal equilibrium, the rate of water feed should match the rate of steam consumption. However the rate of steam consumption varies with speed and gradient, so the fireman will have to adjust the injector settings to suit.

It must be borne in mind that, because of a boiler's length, the apparent water level shown in the gauge glasses varies dependant on whether the locomotive is going uphill, on the level or downhill. Allowance must be made for this. Don't overfill the boiler otherwise water may be carried down into the cylinders. NEVER let the water level drop too low. If the fusible plug melts, high pressure steam will enter the firebox to put the fire out.

6.4.6. Steaming Problems

One of the biggest problem areas with running a steam locomotive is keeping it steaming well. As noted previously, if the fireman can provide the right quantity of steam to meet demand, all is well. However, there are a number of factors which can prevent this from happening.

Firstly, steam locomotives need the right coal. It must remain in lumps as it burns, and not melt like tar, or shatter in the heat. It needs the right calorific value, and preferably a low ash content. Some grades of anthracite have too high a calorific value and will quickly melt the grate. The ash needs to have a high fusion temperature, so that it doesn't fuse into sheets of clinker that will block the airflow through the grate. Oil firing is more expensive, but gives much more consistent results.

Having got the right coal, the skill of the fireman comes into play. The fire needs to be the right thickness, to give the best results for a particular coal. The shape of the fire can influence how it performs - it may need to be saucer shaped, or piled higher at the front or back of the firebox. The fireman learns by experience what to look for when the rate of steam production is poor. Ash built up can be shaken down with a rocking grate, or by use of the fire irons. If clinker formation is particularly bad, it may be necessary to stop the train whilst the fireman cleans the fire. This cannot be done on the move with the fire burning fiercely.

The fireman must take care to ensure that a 'hole' does not appear in the fire, where the firebars are uncovered. If this happens, the bulk of the air being drawn from under the fire will pass through the hole, rather than the rest of the firebed; the coal around the hole will burn extremely hot for a short time, before burning out, whilst the firebed will be deprived of air, and not burn properly. Unless corrected quickly, a 'hole' will result in falling boiler pressure, possibly clinker and even burnt firebars in its vicinity.

The colour of the exhaust gases leaving the chimney is a good indication of how the fire is performing. Ideally the exhaust should be a light grey. If there is too much coal on the grate, or insufficient combustion air, the smoke will be black, and the gas temperature is not as high as it could be. If there is no apparent smoke, the chances are that there is too much air entering the firebox, and this will actually be having a cooling effect on the boiler.

If the boiler tubes are not kept clean, the heat transfer from fire gases to the boiler water will not be as effective and this can affect the steaming rate.

A leak in the smokebox can reduce the vacuum created within it, and this will lead to a reduction in draught and steam production. Similarly, a full ashpan can also reduce the airflow and hamper the steaming rate.

6.4.7. Sprung or Weighted Points

On some railways, points are sprung or weighted to eliminate the need for a signalbox at remote locations. In the facing direction, these points should always be taken at a low speed to avoid the blades moving as the train passes over them. Weighted points that are set against the train can be trailed through and they will then reset after the last vehicle has passed through. Do not stop on a weighted point as unintentionally rolling back may cause a derailment. Once over the point, if possible, check that the blades have reset correctly.

6.5 Stopping a Train

Slowing down a train, which is, of course, running with steel tyres on steel rails, calls for a great deal more judgement and planning than slowing down a rubber-tyred road vehicle. The relative lack of grip means that it is all too easy to lock up the wheels and simply skid along the rails with no control. The UK news media make a great deal out of the number of times it is reported that a 'Signal has been Passed At Danger' (SPAD), but if the rail is slippery, due to leaves, snow, ice or a light shower of rain, having a SPAD is a real possibility. On a steam railway, a period of hot weather also increases the risk, due to a build up of oil on the railheads. The key to good braking is route knowledge, and being aware of the rail conditions. The regulator must be shut, and the brake application made at some considerable distance from the stopping point. The brakes can be applied quite firmly when running at speed, but must be gradually reduced as speed reduces, as there is an increased risk of skidding. This method eliminates jolting the passengers when the train stops, and enables the complete brake release to be achieved more quickly, so that the driver can restart the train promptly.

On most steam railways, trains are not run at high speeds, and stopping is a relatively gentle affair. Often route knowledge can be used to advantage, power can be shut off at the right moment, so that a combination of gradient and rolling resistance will slow the train at the required place, and a light application of the brakes is all that is needed to bring it to a halt.

Braking technique varies depending on the system in use, but the following covers the basics.

6.5.1. Locomotive Brakes

On old style freight trains, the trucks were only fitted with handbrakes so, when on the move, only the locomotive and guard's-van brakes were available. Even though this type of train only ran at low speeds, the driver still needed to plan his braking well in advance. When the locomotive brakes were applied, the train would bunch up behind the locomotive. The driver could give three short blasts on the locomotive whistle to request the guard to apply his brakes. When the train was very nearly at a stand, the driver needed to release the locomotive brakes to stretch the train out, so there was less jolting on the restart.

The locomotive brakes are also used when moving the locomotive on its own (light engine working).

6.5.2. Train Brakes

On passenger trains, air or vacuum brakes are used. Reducing the level of air pressure or vacuum in the train pipe is proportional to the amount of brake force that will be applied. The brake force should be gradually reduced as the train comes to a standstill in order to avoid stopping with a jolt. The

*Fig. 44
Taking water quickly from an overhead gantry in South Africa.*

air or vacuum must be fully restored after a brake application to replenish the brake reservoirs.

6.5.3. Handbrakes

These are generally used when the locomotive is to remain stationary for some while, or is to be left without the crew on the footplate.

6.6 Locomotive Care during the Day

During the course of the day it will be necessary to:-

6.6.1. Take Water

Check that the water tank always has a good supply of water. Familiarity with the railway will allow better judgement of how often this should be filled, and how long it will take.

6.6.2. Lubricate the Motion

The valve gear will need lubricating with motion oil, oilboxes may need to be refilled, and grease filled bearings replenished. Familiarity with a particular locomotive will allow the time interval or mileage between lubrication applications to be judged.

6.6.3. Check Everything is Working

Look to see that the oil is flowing through the cylinder lubricators - most have sight glasses where the oil flow can be seen. Also make sure that the air compressor is creating enough air pressure if air brakes are being used. Listen to the exhaust beat from the chimney.... is it even and regular? If something is not performing correctly, find out why as soon as possible. If it is something you don't understand, get advice. Never ignore a fault. Always report any problems on the locomotive, or on the railway, to the relevant person, so that they can take action as required.

6.6.4. Keep it Clean

The locomotive will do its best to dirty itself during the day. The footplate area should be kept clean from oil spillages and loose coal as these present hazards to the crew. Wipe round to get rid of any dirty water or soot that has settled on the paintwork; this will not make the engine safer, but it will look better, and take less time to dispose of when its duties are over for the day.

6.7 Disposal of the Locomotive at the End of the Day

Even though the trains have stopped running, the engineman's work is not over. There are maintenance and cleaning jobs that must be done before the locomotive can be left. This is known as "disposal".

6.7.1. Blow Down

The purpose of the blowdown valve is to remove water and any dissolved salts or solid deposits from the bottom of the boiler. When water is boiled, the steam is carried away and used, but it leaves any impurities behind in the boiler. As more water is injected into the boiler, and more steam used, these impurities gradually become more and more concentrated. The sediment is bad as it increases

*Fig. 45
Taking water slowly, at the end of a British branch line.*

corrosion and erosion, and the salts are bad as they affect the surface tension of the water, and could lead to priming.

To blow down, the boiler needs to be full to the top of the gauge glass and a good fire ready to allow the boiler to be filled up again. Check that the injectors work before blowing down. Check that there is no one standing near the discharge pipe. Advise any bystanders that there will be a loud noise. Open the blowdown valve and observe the water level drop in the water level gauge. When the level has dropped to about quarter of a glass, shut the blowdown valve. The drop in water level can happen remarkably quickly on some locomotives, so pay attention, otherwise expensive damage can be done by letting out too much water.

6.7.2. Fill the Boiler

Before the fire is dropped, it is important to fill the boiler. It is good practise to have sufficient fire that injecting does not drag the boiler pressure down. Injecting large quantities of water without a good fire results in significant boiler pressure loss (and therefore temperature loss) and the rapid cooling of the boiler will cause thermal stresses that shorten the boiler's life. If the locomotive is to be used the next day, continue injecting water until it is just out of sight at the top of the glass.

If the locomotive is to be unused for a long period of time, position it where it is to be stabled, and inject water until the boiler is completely full and no more can be added. By minimising the air space, this reduces the corrosion that will take place. When the boiler is full, the locomotive MUST NOT be moved under its own power, otherwise priming will occur.

6.7.3. Disposal of the Fire and Ash

In order to allow slow cooling of the boiler, hot embers are left in the firebox so they gradually extinguish overnight. The fire should be well burnt through by this stage and will only require raking to cause smaller embers to drop into the ashpan. The larger embers are raked into a pile on the grate. Judgement must be exercised as to the size of fire left in, as it is important that the pressure gradually goes down, and does not start to go up.

Some locomotives have either a drop grate, where a section of grate is lowered to allow the remains of the fire through, or a rocking grate to help dispose of

Fig. 46
Blowing down at the end of the day, a job for which ear-defenders are recommended!.

the fire's remains.

The ashpan is emptied by opening its doors or slides and letting the ash fall into the ashpit. Often the ashpan needs to be raked to clear it. If the resulting ash pile covers the rails, clear it off, as it may make it difficult to move the locomotive with too much ash on the railhead. Once all ash is removed, close the ashpan doors or slides, and make sure they are locked in place. Quite often lineside fires are started by burning embers escaping from the ashpan - its not always sparks from the chimney that are to blame. Make sure the ashpan dampers are all firmly closed.

6.7.4. Empty the Smokebox

It is important to empty the ash from the smokebox, as the acidic ash will accelerate corrosion. Most smokebox doors are held shut by a "dart and crossbar" arrangement. The "dart" has a pair of handles in the middle of the smokebox door to align and tighten a lug that engages in the crossbar inside the

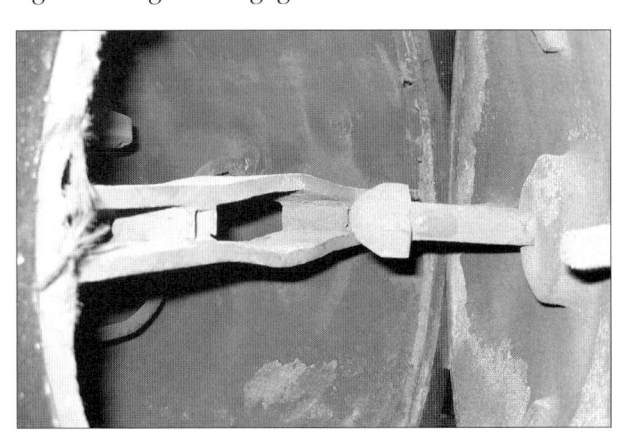

Fig. 47
Smokebox cross bar and dart.

smokebox front. Unscrew (anticlockwise) the outer smokebox door handle until it is felt to be loose. Turn the inner handle through 90 degrees (to point horizontally) and pull the door to open. Observe how ash sits around the door and in the bottom of the smokebox. An uneven pattern in the ash may mean there is an air leak there. It may be necessary to remove the cross bar to get better access. Be careful as it will be very hot.

Brush any ash from the inside of the smokebox, taking care not to damage the fire clay around the pipes. Scoop out the ash and dispose of according to the practises of the railway. Re-fit the cross bar, and close the door. Make sure the inner handle is vertical and the outer handle is tightened sufficiently.

6.7.5. Cap the Chimney

When finished in the smokebox, and when the locomotive is not to be moved again, put a steel plate over the chimney. This is to reduce airflow through the boiler tubes and slow down cooling of the boiler.

6.7.6. Shut Off All Valves

Before leaving the locomotive, ensure all valves are shut. This includes the injector shut off valves, the shut-off to the steam brake and any other auxiliary equipment fitted to the locomotive. Also make sure the regulator is shut, the reverser is in mid gear, the drain cocks open and the handbrake applied. It is also common practice to shut the water level gauge taps to reduce leakage possibilities in the event of a gauge glass breaking whilst the locomotive is unattended.

6.7.7. Drain Air Reservoirs

The air reservoirs on the locomotive accumulate condensation. These should all be fitted with a small drain valve. Give these a short drain at the end of each day if in daily use, or a thorough draining if the locomotive is to be left for a long period of time.

Fig. 48
Locomotive yards and engine sheds are generally dirty - and full of things to trip up the unwary!

Chapter 7
Maintenance and Overhauls

Fig. 49
The tubeplate of a non-superheated locomotive, with a tube-brush just being inserted. Note that the chimney petticoat at top left, and the pipe leading to it, could both also benefit from a good brushing!

Fig. 50
Brushing of the tubes is normally done through the smokebox, but especially if dirty coal is being burnt, a brush at the firebox end, using a short-handled brush, is worth the effort.

7.1 Tube Cleaning

It will be necessary to brush the boiler tubes to clean out any soot or tar deposits that will build up and gradually reduce the ability to steam freely. Typically this is done every three days steaming, but more regularly if the coal type makes it necessary. With the smokebox door open and the cross bar removed, push the round wire brush down the tubes on the end of a suitable push rod. Ensure that it does not become unscrewed by turning the handle clockwise occasionally. If turned anti-clockwise, the brush may come off the handle part way down a tube. Some of the tubes are difficult to get to, so use a more flexible push rod for these. The brush should offer some resistance to being pushed down the tubes, but not be so stiff as to require two people to push it. If the brush offers no resistance at all, it is time to replace it. It is also a good idea to scrape the inside of the chimney to clean off the built up tar-like deposits of burnt cylinder oil.

If a locomotive is worked particularly hard most of the time, the scouring action of the fine ash or char lifted from the firebed will keep the boiler tubes and chimney bore clean enough to make brushing unnecessary.

7.2 Boiler Washout

In order to keep the water in the boiler clean, and reduce the build up of impurities, it is necessary to occasionally drain the water out and refill the boiler. The washout interval may be as little as 6 days of steaming, but with good water treatment could be several weeks. Washout plugs or access holes should be provided along the firebox upper sides, on the backhead, around the firebox foundation ring at the bottom, and also in the front tubeplate. A hosepipe is used to rinse out any deposits and dirt. Any removed washout plugs should be treated with a small amount of special sealant before being replaced and tightened carefully to the right torque. Refill the boiler with treated water before putting in the last top plug.

7.3 Brake Adjustment

As the brake shoes wear, the linkage will need to be adjusted so that they perform correctly. When the shoes wear out completely, these will need to be replaced. This involves removing the clip and pin (or nut and bolt) to allow the old shoe to drop out. The new shoe may require a little fettling to get it to fit. An accumulation of brake dust and oil will restrict brake rigging movement, so the linkage should be cleaned when new shoes are fitted.

7.4 Firebar Replacement

Continuously running with a hot fire will cause the

firebars to melt across their top edge. After a while, the firebars become thinner and may collapse. Firebars can easily be replaced when the boiler is cold by simply lifting out the old bars and placing new ones in.

7.5 Lubrication of Other Parts

Parts that are only occasionally moved, such as the reversing lever or brake rigging, do not need continuous lubrication, but they will benefit from an occasional drop of oil or grease when they are felt to stiffen up. This helps reduce wear and makes them easier to use.

7.6 Overhauls

Any steam fittings and valves should have their sealing packing tightened or replaced as necessary to stop leaks. The valve faces and seats must be ground and lapped to make sure they shut off properly. The boiler must be inspected annually by a qualified inspector, both empty and in steam. This is a legal requirement for all pressure vessels over a certain size. On a less frequent basis, usually seven years, the boiler will have to be removed from the frames, stripped of all fittings, and undergo an ultrasonic and a hydraulic pressure test to ensure its structural integrity. Any failings must be put right otherwise the boiler inspector will not issue a certificate allowing the boiler to be used. At this stage it is normal to replace all the fire tubes, although it may be necessary to replace some of these more frequently.

Whilst the boiler is removed from the frames, the opportunity is usually taken to refurbish the rest of the locomotive.

It will depend on usage, but after a fair mileage has been run, the phosphor bronze bushing in the valve gear and coupling rods will need replacing. Any pins and fittings that look worn should be replaced, and the sealing rings on the main pistons and valves should be checked. Broken or damaged springs should be replaced. Axlebox wear must be taken up to remove any sloppiness otherwise the locomotive begins to ride badly and vibrate when running under power.

Less regularly, the steel tyres on the wheels will need replacing. These are heat shrunk on and very costly. Axleboxes may need re-metalling and re-machining to take up slack. Roller bearings must be checked and replaced as necessary because they can fail catastrophically if left to deteriorate.

Cosmetic restoration is something that should not be neglected otherwise the knocks, chips, dents and scratches soon make the locomotive look untidy.

Not surprisingly, this is all rather expensive!

Fig. 51
Siân during her overhaul in 1999, showing the chassis, smokebox saddle and cylinders.

Fig. 52
Repairing boilers can be fun!
A young volunteer works inside Siân's firebox.

Chapter 8
Driving Miniature and Model Locomotives

*Fig. 53
Models in the smaller gauges - in this case 3½ ", make an excellent introduction to steam locomotive driving for younger enthusiasts.*

All classic steam locomotives work in the same way, whether they run on rails which are 7' or 2½ " apart and, whilst this book concentrates on driving 'big' steam locomotives, everything said so far also applies to driving miniature and model steam locomotives.

However there are a number of specifics, which apply only to driving miniature and model steam locomotives of 15" gauge and under, which should be considered.

8.1 Driver and Fireman

In all these gauges, the driver will also be the fireman; this applies even on 15" gauge steam locomotives where two people can fit in modest comfort on the footplate.

Whilst this means the driver has to be more vigilant of what is happening to his engine, and on the track ahead, miniature and model railways are generally less complex than full-size ones. However, before he starts to fire his locomotive on the move, the driver must double check to ensure that the way ahead is clear, as firing will involve him taking his eyes off the track. For this reason, wherever possible, firing should be carried out whilst stationary.

8.2 Water Supply

In the smaller gauges from 7¼" gauge downwards, locomotive boilers are generally built from copper rather than steel, and care should be taken to ensure as pure a supply of water as possible. Many clubs do provide treated water but if in doubt use distilled water. Unless the locomotive is in regular use, copper boilers should not have water left in them for long periods. Such boilers should be drained via the blow-down valve at low pressure, to ensure the boiler is completely emptied and, when this has been done, the blower and/or injector steam valves should be left open, to admit air as the boiler cools down, and stop any water being sucked in by vacuum from the water tanks.

Finally, and especially in hard-water areas, care should be taken to ensure that the water-ways in gauge glasses, check valves etc. are kept clean, and do not get scaled up.

8.3 Raising Steam

In all these gauges, some form of additional draught is required when lighting the fire, until sufficient boiler pressure is reached to open the blower valve. This can be arranged by an extension chimney in the larger miniature gauges, but for the smaller gauges an electric fan, which can be mounted on the chimney, and which draws the air through from the firebox, is the simplest way - and if battery powered, can be used anywhere. In the smaller gauges, have a good supply of kindling cut to a length to easily fit in the firebox, and keep it in a sealed jar, filled with paraffin. Some charcoal, again soaked in paraffin should be prepared in advance for use after the wood, before switching to dry charcoal, and then

*Fig. 54
The footplate of this 5" gauge GWR 2-6-2 tank locomotive is only 7" wide, but all the major controls found in full-size steam locomotives are there - and work.*

*Fig. 55
On 7¼" gauge, you will be running through points, where extra care is essential.*

coal when the boiler pressure is starting to build up.

8.4 Water Feed

Larger locomotives in 7¼" gauge and upwards will almost always have water feed solely by injector, as described earlier. Smaller locomotives in 7¼" and smaller gauges will almost certainly be fitted with one, and quite possibly two, water feed pumps, as well as an injector. Whilst it is not so true today, historically the reason for pumps being fitted was that injectors do become less reliable the smaller they are, so having a back-up was a necessity.

If only one pump is fitted, it will be a hand pump hidden in the water tank of a tender locomotive, and usually in the rear water tank of a tank locomotive model. A removable handle will protrude through a liftable section of platework for easy operation. Such a pump is very useful for filling the boiler prior to lighting the fire, but considerable care should be taken in using it as a substitute for a temperamental injector - it should only be used for this purpose if the water level in the boiler is known to be above the firebox crown. If there is the slightest doubt about this, the pump must not be used, but the fire dropped and the boiler allowed to cool before water is pumped in. The reason for this is that if there is very little water in the boiler, and there is still a fire in the firebox, pumping cold water into the boiler will severely damage it at best, or cause a violent explosion at worst.

If a second pump is fitted, this will normally work off an eccentric fitted to one of the driving axles. This pump draws water from the water tank and will then pump the water back into the water tank, unless a screw-valve (the "by-pass") in the return pipe is closed, in which case the water is diverted down a branch pipe and through a check valve into the boiler.

Axle pumps are very reliable but suffer (a) because they only work when the locomotive is moving, (b) they use quite a lot of the power developed by the locomotive to pump the water into the boiler against boiler pressure and (c) because they pump in cold water, they reduce the temperature of the water in the boiler, and hence boiler pressure, very quickly; in other words they are inefficient and should only be used if the injector is playing-up.

8.5 Everything Happens More Quickly!

If you open the regulator of a full-size steam locomotive, there is an appreciable pause until the locomotive moves - open the regulator on a 3½" gauge model of that locomotive, and you will move immediately.

Similarly, if you stop the same full size locomotive and don't put the blower on, you will be able to revive the fire a considerable time later - on the model, do the same thing and you will be lucky to revive the fire after a couple of minutes.

Quirks like these are all part of the fun of driving model and miniature locomotives, but they are well worth bearing in mind.

*Fig. 56
Driving a 5" gauge locomotive hauling a dozen passengers takes considerable skill.*

APPENDIX:
SIÂN & THE SIÂN PROJECT GROUP

The Locomotive

Siân was built for the Fairbourne Railway in 1963 by Trevor Guest of Stourbridge, to a design by Ernest Twining. Guest had built many vehicles for the Fairbourne and was a good friend of John Wilkins who ran the little railway in Wales. Ernest Twining had been a colleague of the famous model engineer W.J. Bassett-Lowke. The locomotive was originally to have been called *Zena*, after John Wilkins' secretary, but when she found out there was to be a public naming ceremony, she refused to have the engine named after her. Instead, the engine was named *Siân*, after the daughter of John Wilkins' friend and solicitor.

Siân, and her sister locomotive *Katie*, were the Fairbourne Railway's main motive power from 1964 to 1984. The railway changed ownership in 1984 and *Katie* was sold on to Haigh Hall near Wigan. *Siân* remained at Fairbourne but was rebuilt into an American outline locomotive and was renamed *Sydney* after the new owner's father. Shortly afterwards she was sold to the Littlecote Railway near Hungerford. In 1991, The Littlecote Railway closed and so *Siân* was up for sale again. The newly formed *Siân* Project Group placed a bid for her but were outbid by the Bure Valley Railway in Norfolk. Knowing that the Bure Valley Railway only needed *Siân* as a stopgap measure until their own steam engines were built, the *Siân* Project Group stayed in contact with them and bought the locomotive from them in 1994.

Siân is a 15" gauge, 2-4-2 tender locomotive, with Twining valve gear. She is 21ft long and weighs around 6 tons in working order. The boiler contains about 45 gallons of water, and the tender holds 135 gallons of water. The boiler has a maximum operating pressure of 180psi, although during boiler inspections it is hydraulically tested up to 1.5 times this pressure. The 5" diameter and 8" stroke

Siân as built, at Ravenglass during a visit to The Ravenglass & Eskdale Railway, in 1976.

Siân in American guise as Sydney at Littlecote in 1990.

cylinders, combined with driving wheels of 20" diameter, give a tractive effort of 1,530 lbf. Although never attained due to railway speed restrictions, it is thought that *Siân* could reach speeds of around 30 mph.

The *Siân* Project Group

The *Siân* Project Group (SPG) which is a sub-group of the RAVENGLASS & ESKDALE RAILWAY PRESERVATION SOCIETY, is a non-profit organisation comprising 23 members who have purchased the steam locomotive *Siân* and restored her to original condition. The group bought *Siân* in 1994 from the Bure Valley Railway in Norfolk. She was in serviceable condition, but had been heavily modified by previous owners. She ran each summer up to 1998 and during each winter various parts were rebuilt. A two year thorough overhaul followed involving rebuilding of the boiler, new bearings, repairs to valves and valve gear, and many new parts, as well as restoration to original appearance. Since returning to service in 2000, she has been very reliable and much admired.

However, running steam engines is expensive. During the course of a year, the Group uses several cans of oil for lubrication, several tons of coal for fuel, not to mention all the Brasso for polishing and grease for bearings, and our annual insurance premiums are considerable. On the maintenance front, a new set of boiler tubes costs in the region of £600, a new axle bearing is £150, a steel tyre about £300, and it would require several days work to completely dismantle the engine to carry out any of these repairs!

Group members contribute a monthly maintenance fee to help with the running costs, but at some time in the future a new boiler will be needed and it won't come cheap!

As a small, narrow gauge locomotive, *Siân* costs much less to keep running than a standard gauge locomotive (a mainline express locomotive's seven year overhaul can cost up to £400,000), but the SPG is a small group, and all profits from the sale of this book will go towards keeping *Siân* in full operational order

The driver experience courses, run by the SPG on *Siân*, which allow participants to sample the whole career of an engineman in the space of one day, also contribute to keeping *Siân* running. These courses are run on a one-to-one basis at a fraction of the cost of becoming an owner.

For more information, please visit our website:-

www.sianprojectgroup.com

Siân, the authors and some SPG members at Windmill Farm Railway.

The Authors

The authors are both professional mechanical engineers, with a long history of involvement with preserved steam railways, and are members of the SPG. Along with other group members, they not only drive *Siân*, but carry out much of the maintenance and engineering work as well.

INDEX
REFERENCES THE SECTION IN WHICH THE TOPIC IS COVERED

Ashpan, 5.1.1, 5.1.2, 6.1.2, 6.7.3
Bearings, 5.5, 6.1.6,
Blastpipe, 5.1.1, 5.4
Blowdown/Blow down valve, 4.1, 6.4.3, 6.7.1
Blower, 4.2.3, 5.1.1, 5.7.10,
Boiler, 4.1, 4.2.1, 5.1.1, 5.1.2, 5.1.3, 6.1
Boiler washout, 7.2
Brakes
    Air brake, 5.6.1, 5.6.3, 5.7.6
    Continuous, 5.6.1
    Handbrake, 5.6.1, 5.7.3, 6.5.3
    Steam Brake, 5.6.1
    Vacuum brake, 4.2.3, 5.6.1, 5.6.2, 5.7.5, 6.5.2
Brick Arch, 5.1.1
Chassis, 2.1, 4.2.2
Chimney, 4.1, 4.2.1, 5.1.1, 5.4, 6.7.5, 7.1
Clack, 4.1, 5.1.3
Cleaner, 3.1
Cleaning, 3.1, 6.1.5,
Clinker, 5.1.2, 6.4.6
Combustion, 5.1.2, 6.4.6
Coupling up, 6.2.9
Cow catcher, 4.1
Cut-off, 5.3, 6.3.3, 6.4.1
Cylinder, 2.3, 4.1, 4.2.2, 5.2
Cylinder Drains, 4.1, 4.2.3, 5.7.7, 6.2.5
Dampers, 5.1.1, 5.1.2, 6.7.3
Disposal, 6.7
Dome, 4.1, 5.1.1
Driver, 3.3
Driving wheels, 2.1, 4.1
Ernest Twining, 8.2, Appendix
Fire lighting, 3.1, 6.1.4
Firebox, 4.1, 5.1.1, 5.1.2, 6.1.4
Firehole/Firehole door, 4.2.3, 5.1.2, 5.7
Fireman, 3.2, 4.2.1
Footplate, 4.1
Fusible Plug, 5.1.1, 6.1.2, 6.4.5
Gauge (track gauge), 1
Gauges-
    Boiler pressure, 4.2.3,
    Brake pressure, 4.2.3,
    Steam chest pressure, 4.2.3, 6.3.3
    Water level, 4.2.3, 5.1.1, 5.7.9, 6.1.3, 6.2.2

George Stephenson, 1
Grate, 5.1.1, 5.1.2, 6.4.6, 6.7.3
Grease, 5.5, 6.1.6, 6.6.2, 7.5
Heating (steam heat), 5.7.12
Injector, 4.1, 4.2.1, 4.2.3, 5.1.3, 5.7.8, 6.2.3
Loading Gauge, 1
Lubrication, 5.5, 6.1.6, 6.6.2, 7.5
Lubricator, 4.1, 5.5, 6.1.6, 6.6.3
Main steam pipe, 4.2.1, 4.2.2, 5.1.1, 5.2,
Maintenance, 7, Appendix
Oil
    Bearing oil, 6.1.6, 6.6.2
    Steam oil, 6.1.6
Overhaul, 7.6, Appendix
Piston/piston rod, 4.1, 4.2.2, 5.2, 5.3
Priming, 6.4.3, 6.7.1, 6.7.2
Pony truck, 4.1
Raising steam, 6.1
Regulator, 4.2.1, 5.1.1, 5.7.1, 6.4.1
Reservoir (air), 4.2.3, 5.6.3, 5.7.6, 6.7.7
Reverser, 4.2.2, 4.2.3, 5.3, 5.7.2, 6.4.1
Safety valves, 4.1, 5.1.1
Sanders/sanding, 5.7.11, 6.4.2
Siân Project Group, Appendix
Slipping, 6.3.3, 6.4.2
Smokebox, 4.1, 4.2.2, 5.1.1, 5.4, 6.1.2, 6.4.6, 6.7.4
Smoke deflectors, 5.4.1
Steam chest, 4.2.2, 4.2.3, 5.2, 6.3.3
Steam heating, 5.7.12

Superheat, 5.1.1, 5.5,
Tubes, 5.1.1, 6.4.6, 7.1, 7.6
Turntables, 6.2.8
Valves, 4.2.2, 5.2, 5.3, 7.6
Valve gear, 4.1, 5.3, 6.6.2, 7.6
Valve gears-
    Joy, 5.3.3
    Stephenson, 5.3.1
    Twining, 5.3.4
    Walschaert, 5.3.2
Weighted points, 6.4.7
Wheel arrangements, 2.1,
Whistle, 4.1, 5.1.1, 6.2.6, 6.3.2, 6.5.1

Acknowledgements:

The illustrations in this book were provided by the authors, and by the following to whom all due thanks are rendered.

Drawings and diagrams:-
Terry Taylor and various long-forgotten draughtsmen working in railway company offices many years ago.

Photographs:-

Geoff Asplin
Dave Bedding
Marina Buck
Stan Buck
Olav Casander
John Fuller
Phil Girdlestone
Adam Harris
Gill Holland
Jan Kieboom
Shaun McMahon
Chris Mounsey
Terry Taylor
Dave Wallace